Brian Flem... Y0-DID-841
Ministry of Education
Ministry of Training, Colleges & Universities
900 Bay St. 13th Floor, Mowat Block
Toronto, ON M7A 1L2

Learning Mathematics for Life

A PERSPECTIVE FROM PISA

Programme for International Student Assessment

ORGANISATION FOR ECONOMIC CO-OPERATION AND DEVELOPMENT

The OECD is a unique forum where the governments of 30 democracies work together to address the economic, social and environmental challenges of globalisation. The OECD is also at the forefront of efforts to understand and to help governments respond to new developments and concerns, such as corporate governance, the information economy and the challenges of an ageing population. The Organisation provides a setting where governments can compare policy experiences, seek answers to common problems, identify good practice and work to co-ordinate domestic and international policies.

The OECD member countries are: Australia, Austria, Belgium, Canada, the Czech Republic, Denmark, Finland, France, Germany, Greece, Hungary, Iceland, Ireland, Italy, Japan, Korea, Luxembourg, Mexico, the Netherlands, New Zealand, Norway, Poland, Portugal, the Slovak Republic, Spain, Sweden, Switzerland, Turkey, the United Kingdom and the United States. The Commission of the European Communities takes part in the work of the OECD.

OECD Publishing disseminates widely the results of the Organisation's statistics gathering and research on economic, social and environmental issues, as well as the conventions, guidelines and standards agreed by its members.

> *This work is published on the responsibility of the Secretary-General of the OECD. The opinions expressed and arguments employed herein do not necessarily reflect the official views of the Organisation or of the governments of its member countries.*

ISBN 978-92-64-07499-6 (print)
ISBN 978-92-64-07500-9 (PDF)

Corrigenda to OECD publications may be found on line at: *www.oecd.org/publishing/corrigenda*.
PISA[TM], OECD/PISA[TM] and the PISA logo are trademaks of the Organisation for Economic Co-operation and Development (OECD). All use of OECD trademarks is prohibited without written permission from the OECD.
© OECD 2009

You can copy, download or print OECD content for your own use, and you can include excerpts from OECD publications, databases and multimedia products in your own documents, presentations, blogs, websites and teaching materials, provided that suitable acknowledgment of OECD as source and copyright owner is given. All requests for public or commercial use and translation rights should be submitted to *rights@oecd.org*. Requests for permission to photocopy portions of this material for public or commercial use shall be addressed directly to the Copyright Clearance Center (CCC) at *info@copyright.com* or the Centre français d'exploitation du droit de copie (CFC) at *contact@cfcopies.com*.

Foreword

The recognition of the Programme for International Student Assessment (PISA) in many countries has fostered an interest in the tests the students take. This publication examines the link between the PISA test requirements and student performance. Focus is placed on the proportions of students who answer questions correctly across the range of difficulty from easy, to moderately difficult to difficult. The questions are classified by content, competencies, context and format and analysed to see what connections exist.

This analysis has been carried out in an effort to link PISA results to curricular programmes and structures in participating countries and economies. Results from the assessment reflect differences in country performance in terms of content, competencies, context, and format of the test questions. These findings are important for curriculum planners, policy makers and in particular teachers – especially mathematics teachers of intermediate and lower secondary school classes.

This thematic report is the product of a collaborative effort between the countries involved in PISA, the members of the Mathematics Expert Group listed in Annex A6 who worked to develop the assessment items, the experts who guided the thematic report to its initial form (Jan de Lange, Alla Routitsky, Kaye Stacey, Ross Turner and Margaret Wu), the OECD Directorate for Education staff (principally Andreas Schleicher, Claire Shewbridge and Pablo Zoido with the collaboration of Niccolina Clements), and John Dossey who edited the report in its final form. Juliet Evans provided administrative support and Peter Vogelpoel did the typesetting.

The development of this thematic report was steered by the PISA Governing Board, which is chaired by Lorna Bertrand (United Kingdom). This report is published on the responsibility of the Secretary-General of the OECD.

Lorna Bertrand
Chair of the PISA Governing Board

Barbara Ischinger
Director for Education, OECD

Table of Contents

CHAPTER 1
PISA 2003: INTRODUCTION 11

Introduction 12
Purpose 12
Background 12
Organisation of the report 13

READER'S GUIDE 15

Abbreviations used in this report 15
Technical definitions 16
Further documentation 16

CHAPTER 2
MAIN FEATURES OF THE PISA MATHEMATICS THEORETICAL FRAMEWORK 17

Introduction 18
Mathematical literacy 19
Mathematical content in PISA – the use of overarching ideas 21
- Change and relationships 22
- Space and shape 24
- Quantity 25
- Uncertainty 26

Overarching ideas and traditional topics 27
Context – setting the mathematical problem to be solved 28
- Variety of contexts 29
- PISA contexts 30
- Mathematical relevance of context 30

The competencies 31
- Mathematical thinking and reasoning 32
- Mathematical argumentation 32
- Modelling 32
- Problem posing and solving 32
- Representation 33
- Symbols and formalism 33
- Communication 33
- Aids and tools 33

Learning Mathematics for Life: A Perspective from PISA – © OECD 2009

Table of Contents

Competency clusters. 34
- Reproduction cluster. 34
- Connections cluster. 35
- Reflection cluster. 36

Conclusion 36

CHAPTER 3
A QUESTION OF DIFFICULTY: QUESTIONS FROM PISA 2003. 39

Introduction 40

Describing growth in *mathematical literacy*: how difficult is the question and
where does it fit on the PISA mathematics scale? 40

The PISA scale and difficulty 41

Examples of the easiest mathematics questions from PISA 2003 46

Examples of moderate to difficult mathematics questions from PISA 2003 62

Examples of difficult mathematics questions from PISA 2003 96

Conclusion 116

CHAPTER 4
COMPARISON OF COUNTRY LEVEL RESULTS 117

Introduction 118

Cross country differences in curriculum 119

Groupings of countries by patterns in item responses 120

Patterns in mathematics content 122

Performance and grade levels 129

Competency clusters and mathematics performance. 133

Context and mathematics performance 135

Conclusion 137

CHAPTER 5
THE ROLES OF LANGUAGE AND ITEM FORMATS. 139

Introduction 140

The use of language in PISA mathematics questions and student performance 140

Word-count and question difficulty across countries 141

Word-count and the context in which a question is presented 143

Word-count and competencies required to answer the question 146

Word-count and content 147

Item-format and mathematics performance 149

Item-format and item difficulty across countries 150

Item-format, the three C's and word-count 151

Differences in item-format and omission rates 152

Conclusion 154

CHAPTER 6
MATHEMATICAL PROBLEM SOLVING AND DIFFERENCES IN STUDENTS' UNDERSTANDING 157

Introduction 158
General features of mathematical problem solving in PISA 158
Making the problem-solving cycle visible through case studies of questions 160
- The first case study: Bookshelves – Question 1 161
- Reflection on Bookshelves – Question 1 163
- The second case study: Skateboard – Question 3 164
- Reflection on Skateboard – Question 3 167

Students' mathematical understandings and item scoring 167
- Item coding in the database and information on students' thinking 167

Students' understanding of proportional reasoning 177
- The prevalence of proportional reasoning in PISA questions 177
- The difficulty of proportional reasoning questions 178

Students' understanding of symbolic algebra 182
Students' understanding of average 185
Conclusion 187

REFERENCES 189

ANNEX A1
PISA 2003 MATHEMATICS ASSESSMENTS:
CHARACTERISTICS OF QUESTIONS USED 193

ANNEX A2
OTHER EXAMPLES OF PISA MATHEMATICS QUESTIONS THAT WERE
NOT USED IN THE PISA 2003 MATHEMATICS ASSESSMENT 231

ANNEX A3
TRADITIONAL DOMAINS AND PISA ITEMS 235

ANNEX A4
WORD-COUNT AND THE 3 Cs – ANALYSIS OF VARIANCE 237

ANNEX A5
ANALYSIS OF VARIANCE RELATED TO ITEM FORMAT 241

ANNEX A6
MATHEMATICS EXPERT GROUP 245

Figures

Figure 2.1	Components of the PISA mathematics domain	18
Figure 3.1a	PISA mathematics proficiency Levels 1 and 2: Competencies students typically show and publicly released questions	42
Figure 3.1b	PISA mathematics proficiency Levels 3 and 4: Competencies students typically show and publicly released questions	43
Figure 3.1c	PISA mathematics proficiency Levels 5 and 6: Competencies students typically show and publicly released questions	44
Figure 4.1	Comparison of item parameters by countries for three selected items	121
Figure 4.2	Hierarchical cluster analysis of item parameters	123
Figure 4.3	Relative difficulty by mathematics topic by country	124
Figure 4.4	Average performance by grade for four participants	130
Figure 4.5	Average question difficulty by competency cluster in participating countries	133
Figure 4.6	Average question difficulty by competency cluster and by traditional topic	134
Figure 4.7	Average question difficulty by context in participating countries	135
Figure 5.1	Average relative difficulty of questions within each word-count category for each country	142
Figure 5.2	Context and length of question by average relative difficulty of questions	144
Figure 5.3	Competency clusters and length of question by average relative difficulty of questions	146
Figure 5.4	Traditional mathematics topics and length of question by average relative difficulty of questions	148
Figure 5.5	Average item difficulty (logits) by item-format by country	150
Figure 5.6	Average relative difficulty of questions by item-format and word-count	152
Figure 6.1	Mathematisation cycle	159
Figure 6.2	Performance of some English speaking countries on proportional reasoning items, illustrating their similar pattern of performance	179
Figure 6.3	Proportional reasoning performances of Austria and Sweden, showing variation in Exchange Rate – Question 3	182
Figure 6.4	Performance on algebra items for countries scoring highly on the content items from *change and relationships*	183
Figure 6.5	Performance on algebra items for the countries scoring at the OECD average on the content items from *change and relationships*	184
Figure 6.6	Results of selected countries on HEIGHT concerning the mathematical concept of average	185
Figure 6.7	Results of selected countries on some non-released items concerning the mathematical concept of average	186
Figure A1.1	Student performance on Exchange Rate – Question 1	200
Figure A1.2	Student performance on Staircase – Question 1	201
Figure A1.3	Student performance on Exports – Question 1	202
Figure A1.4	Student performance on Exchange Rate – Question 2	203
Figure A1.5	Student performance on The Best Car – Question 1	204
Figure A1.6	Student performance on Growing Up – Question 1	205
Figure A1.7	Student performance on Growing Up – Question 2	206
Figure A1.8	Student performance on Cubes – Question 1	207
Figure A1.9	Student performance on Step Pattern – Question 1	208
Figure A1.10	Student performance on Skateboard – Question 1	209
Figure A1.11	Student performance on Bookshelves – Question 1	210
Figure A1.12	Student performance on Number Cubes – Question 2	211
Figure A1.13	Student performance on Internet Relay Chat – Question 1	212
Figure A1.14	Student performance on Coloured Candies – Question 1	213
Figure A1.15	Student performance on Litter – Question 1	214
Figure A1.16	Student performance on Skateboard – Question 3	215
Figure A1.17	Student performance on Science Tests – Question 1	216
Figure A1.18	Student performance on Earthquake – Question 1	217
Figure A1.19	Student performance on Choices – Question 1	218
Figure A1.20	Student performance on Exports – Question 2	219

Figure A1.21	Student performance on Skateboard – Question 2	220
Figure A1.22	Student performance on Growing Up – Question 3	221
Figure A1.23	Student performance on Exchange Rate – Question 3	222
Figure A1.24	Student performance on P2000 Walking – Question 1	223
Figure A1.25	Student performance on Support for the President – Question 1	224
Figure A1.26	Student performance on Test Scores – Question 1	225
Figure A1.27	Student performance on Robberies – Question 1	226
Figure A1.28	Student performance on Internet Relay Chat – Question 2	227
Figure A1.29	Student performance on The Best Car – Question 2	228
Figure A1.30	Student performance on Walking – Question 3	229
Figure A1.31	Student performance on Carpenter – Question 1	230

Tables

Table 2.1	Examples of *change and relationships* questions	23
Table 2.2	Examples of *space and shape* questions	24
Table 2.3	Examples of *quantity* questions	26
Table 2.4	Examples of *uncertainty* questions	27
Table 2.5	Cross-tabulation of PISA items by PISA and traditional topics classifications	28
Table 2.6	Examples of questions in the *reproduction* competency cluster	34
Table 2.7	Examples of questions in the *connections* competency cluster	35
Table 2.8	Examples of questions in the *reflection* competency cluster	36
Table 3.1	Characteristics of the easiest released PISA 2003 mathematics questions	47
Table 3.2	Characteristics of moderate to difficult released questions	63
Table 3.3	Characteristics of the most difficult questions released from the PISA 2003 mathematics assessment	97
Table 4.1	Mean and standard deviation of relative topic difficulty across countries	125
Table 4.2	Relative easiness/difficulty of each topic within the countries	126
Table 4.3	Average item difficulty parameter values for Data items[1]	128
Table 4.4	Items identified with grade DIF for countries with multiple grades	131
Table 4.5	Mean and standard deviation of question difficulty by competency cluster across countries	133
Table 4.6	Questions in competency clusters by traditional mathematics topic	134
Table 4.7	Mean and standard deviation of question difficulty by context across countries	136
Table 4.8	Multiple comparisons of question difficulty by context across countries (using Bonferroni adjustment)	136
Table 5.1	Mean and standard deviation of difficulty of questions in each word-count category across countries	143
Table 5.2	Item distribution by context by word-count	145
Table 5.3	Average number of words by context by word-count	145
Table 5.4	Item distribution by competencies by word-count	147
Table 5.5	Distribution of questions by traditional topic and length of question	149
Table 5.6	Mean and standard deviation of item difficulty in item-format categories across countries	151
Table 5.7	Average percent of missing data by item difficulty for three item-format categories – PISA Field Trial 2003	154
Table 6.1	Use of different types of PISA 2003 mathematics question formats	168
Table 6.2	Distribution of responses for Exports – Question 2	171
Table 6.3	Examples of multiple-choice questions in Chapter 3	171
Table 6.4	Examples of questions with double-digit coding in Chapter 3	176
Table 6.5	Instances of proportional reasoning in questions presented in Chapter 3	177
Table 6.6	Hierarchy of proportional reasoning items (Hart, 1981)	178
Table 6.7	Level of difficulty of proportional reasoning questions (PISA proficiency level, question difficulty parameter, Hart level)	179

Table A1.1	Characteristics of released PISA 2003 mathematics items	194
Table A2.1	Other examples of released PISA mathematics questions not used in PISA 2003	232
Table A3.1	Traditional domains; average item difficulties (logits) relative to other topics and their standard deviations	236
Table A4.1	Full factorial ANOVA with word-count and country as factors	238
Table A4.2	Full factorial ANOVA with word-count, country and competencies as factors	238
Table A4.3	Full factorial ANOVA with word-count, country and content as factors	238
Table A4.4	Full factorial ANOVA with word-count, country and context as factors	239
Table A4.5	Post hoc comparisons for word-count mean difficulties using Bonferroni adjustment	239
Table A5.1	Full factorial ANOVA with item-format and country as factors	242
Table A5.2	Full factorial ANOVA with item-format, country and competencies as factors	242
Table A5.3	Full factorial ANOVA with item-format, country and context as factors	242
Table A5.4	Full factorial ANOVA with item-format, country and word-count as factors	243
Table A5.5	Item distribution across item-format categories and traditional topics	243
Table A5.6	Post hoc comparisons for item format mean difficulties using Bonferroni adjustment	244

PISA 2003: Introduction

The present study affords an opportunity to view 15-year-old students' capabilities internationally through the lens of mathematical literacy *as defined by the PISA 2003 mathematics framework and the resulting assessment. The framework (Chapter 2), the focus on the actual items (Chapter 3), students' performance by mathematical subtopic areas and competency clusters (Chapter 4), the influence of item format and reading level on item difficulty (Chapter 5), and the assessment and interpretation of student problem solving (Chapter 6) present an interesting view of* mathematical literacy *and instruction in an international context.*

INTRODUCTION

This chapter provides an overview of the purposes and goals of this report. It links the important findings of the PISA 2003 mathematics assessment with ways in which they can be put to practical use by teachers in classrooms and by policy makers involved with matters related to instructional practices in mathematics classrooms. In doing so, the report highlights the importance of a focus on *mathematical literacy*, as defined by the PISA programme, to educational programmes worldwide.

PURPOSE

How is mathematical literacy related to curriculum and instruction across countries?

The objective of this report is to provide information that relates the results of the PISA 2003 assessment of *mathematical literacy* to mathematics instruction. Specific focus is given to the exploration of connections between the results obtained, on the one hand, and instructional practices, curriculum, assessment practices, students' problem solving methods, and mathematical thinking on the other hand.

By using the term "literacy", the PISA framework[1] emphasises that mathematical knowledge and skills that have been defined within traditional school mathematics curricula are not the primary focus of the study. Instead, PISA focuses on students' mathematical knowledge as it is put to functional use in varied contexts and in reflective ways which may require insight and some creativity. However, such uses of mathematics are based on knowledge and skills learned in and practised through the kinds of problems that appear in school textbooks and classrooms. Internationally, educational systems have different curricula that result in different emphases placed on applications, different expectations for the use of mathematical rigor and language and different teaching and assessment practices.

The examination of the results related to *mathematical literacy* from PISA 2003 across participating countries makes it possible to identify some associations between the related levels of achievement and instructional practices found within these countries. Such information will be of direct interest to a wide community of educators including teachers, curriculum developers, assessment specialists, researchers, and policy makers.

BACKGROUND

The Programme for International Student Assessment (PISA) is a project of the Organisation for Economic Co-operation and Development (OECD). PISA is a collaborative activity among the 30 member countries of the OECD and some partner countries and economies, bringing together scientific expertise from the participating countries and steered jointly by their governments through a Board, on the basis of shared, policy-driven interests. The project is implemented by a consortium of international researchers led by the Australian Council for Educational Research (ACER).

---

1. The PISA 2003 Assessment Framework (OECD, 2003) is described in detail in Chapter 2.

PISA involves testing of literacy in reading, mathematics, and science in samples of 15-year-olds draw from each participating country. The aim in focusing on students of this age is the generation of a summative, comparative, international report on *mathematical literacy* for students nearing the end of their period of compulsory schooling. The tests are designed to generate measures of the extent to which students can make effective use of what they have learned in school to deal with various problems and challenges they are likely to experience in everyday life. The tests, common across all countries, are translated into the local instructional languages used in each country. Testing first took place in 2000, when reading in the language of instruction was the major test domain. The second cycle of testing occurred in 2003, with *mathematical literacy* the major test domain. The third cycle of testing occurred in 2006, with *scientific literacy* as the major domain focus. PISA collects assessment data every three years with the three domains rotating as the major focus of interest and smaller portions of the assessments being focused on the two other domains. As a result, this OECD programme provides trend data focused on the domains for the participating countries.

PISA seeks to assess how well 15-year-olds are prepared for life's challenges…

… and assesses students in three different domains: reading, mathematics and science.

A typical test cycle has a number of phases – establishment or refinement of the domain frameworks and sample indicators upon which the assessment will focus, development of assessment instruments linked to these frameworks, field trials of all resulting test instruments in all of the participating countries, careful refinement of the assessments and school and student sampling based on these field trials, implementation of the main study in sampled schools from the participating nations, careful cleaning and analysis of the resulting data, and, finally, interpretation and reporting of the results. The PISA assessments for 2000, 2003, and 2006 have resulted in various publications, including the frameworks (OECD, 1999, 2003), initial reports (OECD, 2001, 2004a, 2004b), associated technical reports (OECD, 2002, 2005, 2009a), a number of thematic reports like this one (OECD, 2009b, 2009c, and 2009d) and a wide variety of national level reports (see *www.pisa.oecd.org* for many examples).

ORGANISATION OF THE REPORT

This report concentrates on in-depth analysis of PISA 2003 mathematics performance data at the level of individual tasks and test items.

Chapter 2 provides a detailed description of PISA 2003 assessment framework (OECD, 2003). It explains in detail the constructs of the mathematics assessment in PISA and lays out the context for the examples and further analysis presented in subsequent chapters.

Chapter 3 illustrates this framework with released assessment items and links them to different levels of *mathematical literacy* proficiency. The reader can find the actual items in this chapter along with a discussion of students' performance on each of them.

Chapter 4 focuses on differences in the patterns of performance by aspects of mathematical content contained within the items' expectations. In participating countries, by the age of 15 students have been taught different subtopics from the broad mathematics curriculum and these subtopics have been presented to them differently depending on the instructional traditions of the country.

Chapter 5 focuses on factors other than the three Cs (mathematical content, competencies and context) which influence students' performances. Just as countries differ, students' experiences differ by their individual capabilities, the instructional practices they have experienced, and their everyday lives.

For example, item format, wording, reading demand, the amount of information as well as the use of graphics and formulae in items, can all affect students' performance. Chapter 5 examines some of these differences in the patterns of performance by focusing on three factors accessible through data from PISA 2003: language structure within items, item format, and student omission rates related to items.

The final chapter, Chapter 6, concentrates on problem solving methods and differences in students' mathematical thinking. The PISA 2003 assessment framework (OECD, 2003) gives rise to further possibilities for investigating fundamentally important mathematical problem solving methods and approaches. In particular, the framework discusses processes involved with what is referred to as the "mathematisation" cycle. This incorporates both horizontal mathematisation, where students must link phenomena in the real world with the mathematical world (the emphasis is on creating mathematical models, and on interpretation of real situations in relation to their mathematical elements, or interpreting mathematical representations in relation to their real-world implications), and perform vertical mathematisation, where students are required to apply their mathematical skills to link and process information and produce mathematical solutions. The chapter provides two case studies, explaining how the elements required in the different stages of mathematisation are implemented in PISA items.

Reader's Guide

ABBREVIATIONS USED IN THIS REPORT

Organisations

The following abbreviations are used in this report:

ACER   Australian Council For Educational Research
OECD   Organisation for Economic Cooperation and Development
PISA   The Programme for International Student Assessment
TCMA   Test-Curriculum Match Analysis
TIMSS   Trends in Mathematics and Science Study

Country codes

OECD Countries

CODE	COUNTRY	CODE	COUNTRY
AUS	Australia	MEX	Mexico
AUT	Austria	NLD	Netherlands
BEL	Belgium	NZL	New Zealand
CAN	Canada	NOR	Norway
CZE	Czech Republic	POL	Poland
DNK	Denmark	PRT	Portugal
FIN	Finland	KOR	Korea
FRA	France	SVK	Slovak Republic
DEU	Germany	ESP	Spain
GRC	Greece	SWE	Sweden
HUN	Hungary	CHE	Switzerland
ISL	Iceland	TUR	Turkey
IRL	Ireland	GBR	United Kingdom (England, Wales and Northern Ireland)
ITA	Italy		
JPN	Japan	SCO	Scotland
LUX	Luxembourg	USA	United States

OECD Partner Countries and Economies

CODE	COUNTRY	CODE	COUNTRY
BRA	Brazil	PER	Peru
HKG	Hong Kong-China	RUS	Russian Federation
IDN	Indonesia	YUG	Serbia
LVA	Latvia	THA	Thailand
LIE	Liechtenstein[1]	TUN	Tunisia
MAC	Macao-China	URY	Uruguay

1. Liechtenstein's results are not included in results requiring a separate national scaling of item values as the sample size in the country was too small to provide an accurate result.

PISA items and item codes

PISA tests consist of units, which contain a stimulus and one or more items related to the stimulus (see, for example, Annex A1, WALKING). Each of these units has a code (*e.g.* M124). Each item within the unit has its own code (*e.g.* M124Q01, M124Q02). The item names and a question number, *e.g.* WALKING Q1, are used to identify particular items.

Some of the PISA items are secured for future use and cannot be shown in this report. However, a number of PISA mathematics items have been released into the public domain. All released items from PISA 2003 are placed in Annex A1.

TECHNICAL DEFINITIONS

Item difficulty – Historically, item difficulty is the proportion of those taking an item, or test, which get the item correct. Within situations employing item response theory (IRT) modelling of response to items relative to the underlying trait (*e.g. mathematical literacy* in the area being measured), item difficulty is the value on the trait scale where the slope of the item's corresponding item response function reaches its maximal value.

Fifteen-year-olds – The use of fifteen-year-olds in the discussion of the PISA sample population refers to students who were aged between 15 years and 3 (complete) months and 16 years and 2 (complete) months at the beginning of the assessment period and who were enrolled in an educational institutions regardless of grade level or institution type or if they were enrolled as a full-time or part-time students.

OECD average – Takes the OECD countries as single entities, each with equal weight. Hence, an OECD average is a statistic generated by adding the country averages and dividing by the number of OECD countries involved. The OECD average provides data on how countries rank relative to the set of countries within the OECD.

OECD total – Takes the OECD countries merged as a single entity to which each country contributes in proportion to the number of its students in the appropriate population. The computation of the OECD total involves the sum total of the outcome variable of interest divided by the total number of data-related students within the OECD countries. The OECD total provides a comparison statistic for the total human capital present with the OECD countries.

Rounding of numbers – Because of rounding, some columns or groups of numbers may not add up to the totals shown. Totals, differences, and averages are always calculated on the basis of exact numbers and then rounded after calculation.

FURTHER DOCUMENTATION

For further documentation on the PISA assessment instruments and the methods used in PISA, see the PISA 2003 Technical Report (OECD, 2005), the Australian Council of Educational Research PISA site (*www.acer.edu.au/ozpisa*) and the PISA web site (*www.pisa.oecd.org*).

Main Features of the PISA Mathematics Theoretical Framework

This chapter provides a detailed description of the PISA 2003 assessment framework (OECD, 2003). It explains in detail the constructs of the mathematics assessment in PISA and lays out the context for the examples and further analysis presented in subsequent chapters.

INTRODUCTION

In order to appreciate and evaluate the mathematics items used in PISA it is important to understand the theoretical mathematics framework used for the assessment (OECD, 2003). This overview will focus on highlights of the framework, and illustrate these by means of PISA assessment items that have been released into the public domain.

Content, contexts, competencies and mathematical literacy are the building blocks for the PISA mathematics framework.

The structure of the PISA mathematics framework can be characterised by the mathematical representation: ML + 3Cs. ML stands for *mathematical literacy*, and the three Cs stand for content, contexts and competencies. Suppose a problem occurs in a situation in the real world; this situation provides a context for the mathematical task. In order to use mathematics to solve the problem, a student must have a degree of mastery over relevant mathematical content. And in order to solve the problem a solution process has to be developed and followed. To successfully execute these processes, a student needs certain competencies, which the framework discusses in three competency clusters.

This chapter begins with a discussion of *mathematical literacy*, and then outlines the three major components of the mathematics domain: context, content and competencies. These components can be illustrated schematically in Figure 2.1, reproduced from the framework (OECD, 2003).

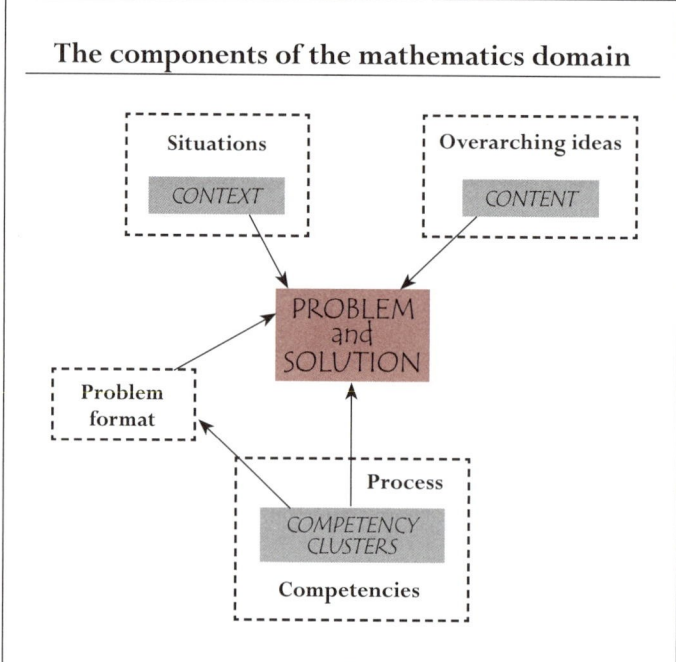

Figure 2.1 ■ **Components of the PISA mathematics domain**

Source: OECD (2004a), Learning for Tomorrow's World: First Results from PISA 2003, OECD Publications, Paris.

MATHEMATICAL LITERACY

The PISA *mathematical literacy* domain is concerned with the capacities of students to analyse, reason, and communicate ideas effectively as they pose, formulate, solve and interpret mathematical problems in a variety of situations. The accompanying assessment focuses on real-world problems, moving beyond the kinds of situations and problems typically encountered in school classrooms. In real-world settings, citizens regularly face situations when shopping, travelling, cooking, dealing with personal finances, analysing political positions, and considering other issues where the use of quantitative or spatial reasoning or other mathematical competencies would be of help in clarifying or solving a problem.

PISA defines a form of mathematical literacy…

Such uses of mathematics are based on knowledge and skills learned and practised through the kinds of problems that typically appear in school textbooks and classrooms. However, these contextualised problems demand the ability to apply relevant skills in a less structured context, where the directions are not so clear for the students. Students have to make decisions about what knowledge may be relevant, what process or processes will lead to a possible solution, and how to reflect on the correctness and usefulness of the answer found.

Citizens in every country are increasingly confronted with a myriad of issues involving quantitative, spatial, probabilistic or relational reasoning. The media are full of information that use and misuse tables, charts, graphs and other visual representations to explain or clarify matters regarding weather, economics, medicine, sports, and environment, to name a few. Even closer to the daily life of every citizen are skills involving reading and interpreting bus or train schedules, understanding energy bills, arranging finances at the bank, economising resources, and making good business decisions, whether it is bartering or finding the best buy.

… that requires engagement with mathematics…

Thus, literacy in mathematics is about the functionality of the mathematics an individual learned at school. This functionality is an important survival skill for the citizen in today's information and knowledge society.

The definition of *mathematical literacy* for PISA is:

> Mathematical literacy *is an individual's capacity to identify, and understand, the role that mathematics plays in the world, to make well-founded judgments and to use and engage with mathematics in ways that meet the needs of that individual's life as a constructive, concerned, and reflective citizen.* (OECD 2003)

Some explanatory remarks are in order for this definition to become transparent.

In using the term "literacy", the PISA focus is on the sum total of mathematical knowledge a 15-year-old is capable of putting into functional use in a variety of contexts. The problems often call for reflective approaches involving insight and some creativity. As such, PISA focuses on the mathematical knowledge and skills that go beyond the mathematics that has been defined within and limited to the outcomes of a school curriculum.

Mathematical literacy cannot be reduced to – but certainly presupposes – knowledge of mathematical terminology, facts and procedures as well as numerous skills in performing certain operations, and carrying out certain methods. PISA emphasises that the term "literacy" is not confined to indicating a basic, minimum level of functionality. On the contrary, PISA considers literacy as a continuous and multi-faceted spectrum ranging from aspects of basic functionality to high-level mastery.

… going beyond the mastery of mathematical techniques conventionally taught at school.

A crucial capacity implied by our notion of *mathematical literacy* is the ability to pose, formulate and solve intra- and extra-mathematical problems within a variety of domains and contexts. These range from purely mathematical ones to ones in which no mathematical structure is present from the outset but may be successfully introduced by the problem poser, problem solver, or both.

Attitudes and emotions (*e.g.* self-confidence, curiosity, feelings of interest and relevance, desire to do or understand things) are not components of the definition of *mathematical literacy*. Nevertheless they are important prerequisites for it. In principle it is possible to possess *mathematical literacy* without possessing such attitudes and emotions at the same time. In practice, however, it is not likely that such literacy will be exerted and put into practice by someone who does not have some degree of self-confidence, curiosity, feeling of interest and relevance, and desire to do or understand things that contain mathematical components.

The concept of *mathematical literacy* is by no means new. Related terms that have been used to describe it have varied from numeracy to quantitative literacy. Historically, Josiah Quincy connected the responsibility of citizens and lawmakers with statistical knowledge in 1816 and called it "political arithmetic". Since the identification of this linkage, attention has been given to the relation between the functionality of mathematics and needs of the responsible citizen. The definition of what constitutes *mathematical literacy* still varies widely from very narrow definitions like "the knowledge and skills required to apply arithmetic operations, either alone or sequentially, using numbers embedded in printed material" to "the ability to cope confidently with the mathematical needs of adult life" (Cohen, 2001).

Mathematical literacy is about dealing with "real" problems. That means that these problems are typically placed in some kind of a "situation". In short, the students have to "solve" a real world problem requiring them to use the skills and competencies they have acquired through schooling and life experiences. A fundamental process in this is referred to as "mathematisation". This process involves students shifting between the real-world context of the problem and the mathematical world needed to solve it. Mathematisation involves students in interpreting and evaluating the problem and reflecting on its solution to ensure that the solution obtained indeed addresses the real situation that engendered the problem initially.

It is in this sense that *mathematical literacy* goes beyond curricular mathematics. Nevertheless, the assessment of *mathematical literacy* can't be separated from existing curricula and instruction because students' knowledge and skills largely depend on what and how they have learnt at school and on how this learning has been assessed. The analysis will continue through a discussion of the three Cs – content, context and competencies.

The process of "mathematisation" describes the ability of students to solve real-world problems by shifting between real-world and mathematical world contexts.

MATHEMATICAL CONTENT IN PISA – THE USE OF OVERARCHING IDEAS

Mathematics school curricula are typically organised into topics and place an emphasis on procedures and formulas. This organisation sometimes makes it difficult for students to see or experience mathematics as a continuously growing scientific field that is constantly spreading into new fields and applications. Students are not positioned to see overarching concepts and relations, so mathematics appears to be a collection of fragmented pieces of factual knowledge.

"What is mathematics?" is not a simple question to answer. A person asked at random will most likely answer, "Mathematics is the study of numbers." Or, perhaps, "Mathematics is the science of numbers." And, as Devlin (1997) states in his book *Mathematics: The Science of Patterns*, the former is a huge misconception based on a description of mathematics that ceased to be accurate some 2 500 years ago. Present-day mathematics is a thriving, worldwide activity; it is an essential tool for many other domains like banking, engineering, manufacturing, medicine, social science, and physics. The explosion of mathematical activity that has taken place in the twentieth century has been dramatic.

At the turn of the nineteenth century, mathematics could reasonably be regarded as consisting of about a dozen distinct subjects: arithmetic, algebra, geometry, probability, calculus, topology, and so on. The typical present-day school curricula topics are drawn from this list.

A more reasonable figure for today, however, would be 70 to 80 distinct subjects. Some subjects (*e.g.* algebra, topology) have split into various sub fields; others (*e.g.* complexity theory, dynamical systems theory) are completely new areas of study (see, for example, the American Mathematical Society's *Mathematics by Classification*, 2009).

Mathematics can be seen as a language that describes patterns: patterns in nature and patterns invented by the human mind. Those patterns can either be real or imagined, static or dynamic, qualitative or quantitative, purely utilitarian or of little more than recreational interest. They can arise from the world around us, from the depth of space and time, or from the inner workings of the human mind.

PISA organises mathematical content into four overarching ideas.

PISA aims to assess students' capacity to solve real problems, and therefore includes a range of mathematical content that is structured around different phenomena describing mathematical concepts, structures or ideas. This means describing mathematical content in relation to the phenomena and the kinds of problems for which it was created. In PISA these phenomena are called "overarching ideas". Using this approach PISA also covers a range of mathematical content that includes what is typically found in other mathematics assessments and in national mathematics curricula. However, PISA seeks to assess whether students can delve deeper to find the concepts that underlie all mathematics and therefore demonstrate a better understanding of the significance of these concepts in the world (for more information on phenomenological organisation of mathematical content see Steen, 1990).

The domain of mathematics is so rich and varied that it would not be possible to identify an exhaustive list of related phenomenological categories. PISA assesses four main overarching ideas:

- *Change and relationships*
- *Space and shape*
- *Quantity*
- *Uncertainty*

These four overarching ideas ensure the assessment of a sufficient variety and depth of mathematical content and demonstrate how phenomenological categories relate to more traditional strands of mathematical content.

Change and relationships

Change and relationships involves the knowledge of mathematical manifestations of change, as well as functional relationships and dependency among variables.

PISA recognises the importance of the understanding of *change and relationships* in *mathematical literacy*. Every natural phenomenon is a manifestation of change. Some examples are organisms changing as they grow, the cycle of seasons, the ebb and flow of tides, cycles for unemployment, weather changes, and the Dow-Jones index. Some of these change processes can be described or modelled by some rather straightforward mathematical functions (*e.g.* linear, exponential, periodic, logistic, either discrete or continuous). But many processes fall into different categories, and data analysis is quite often essential. The use of computer technology has resulted in more powerful approximation techniques, and more sophisticated visualisation of data. The patterns of change in nature and in mathematics do not in any sense follow the traditional mathematical content strands.

Following Stewart (1990), PISA is sensitive to the patterns of change and aims to assess how well students can:

- represent changes in a comprehensible form;

- understand the fundamental types of change;

- recognise particular types of changes when they occur;

- apply these techniques to the outside world and

- control a changing universe to our best advantage.

The PISA overarching ideas of *change and relationships* includes many different traditional topics, most obviously functions and their representations, but also series. Further, *change and relationships*, as an overarching idea, encompasses patterns occurring in nature, art, and architecture in geometric situations.

Table 2.1 lists all of the released *change and relationships* questions that were used in the main PISA mathematics assessment and where readers can find these in Chapter 3. For example, the unit GROWING UP presents students with a graph showing the functional relationship between height in centimetres and age in years for a particular group of young males and young females. Question 1 invites students to interpret a statement about growth (change in height) over time, then to identify and carry out a simple calculation. Question 2 asks students to interpret the graph to identify the time period in which a certain relationship exists between heights of the females and males. Question 3 invites students to explain how the graph shows an aspect of change in growth rate.

Table 2.1
Examples of *change and relationships* questions

Question	Where to find question in Chapter 3	
THE BEST CAR	Question 1	
GROWING UP	Question 1	Examples of easy questions section
GROWING UP	Question 2	
INTERNET RELAY CHAT	Question 1	Examples of questions of moderate difficulty section
GROWING UP	Question 3	
WALKING	Question 1	
INTERNET RELAY CHAT	Question 2	Examples of difficult questions section
THE BEST CAR	Question 2	
WALKING	Question 3	

Space and shape

PISA recognises that patterns are encountered not only in processes of *change and relationships*, but also can be explored in a static situation. Shapes are patterns: houses, churches, bridges, starfish, snowflakes, city plans, cloverleaves, crystals, and shadows. Geometric patterns can serve as relatively simple models of many kinds of phenomena, and their study is possible and desirable at all levels. Shape is a vital, growing, and fascinating theme in mathematics that has deep ties to traditional geometry (although relatively little in school geometry) but goes far beyond it in content, meaning, and method.

Space and shape relates to the understanding of spatial and geometric phenomena and relationships.

In the study of shape and constructions, students should look for similarities and differences as they analyse the components of form and recognise shapes in different representations and different dimensions. The study of shapes is closely knitted to "grasping space". That is learning to know, explore, and conquer in order to improve how we live, breathe, and move through the space in which we live (Freudenthal, 1973; Senechal, 1990).

Students must be able to understand relative positions of objects and to be aware of how they see things and why they see them this way. Students must learn to navigate through space and through constructions and shapes. Students should be able to understand the relation between shapes and images or visual representations (*e.g.* the relation between the real city and photographs or maps of the same city). They must also understand how three-dimensional objects can be represented in two dimensions, how shadows are formed and interpreted, and what "perspective" is and how it functions.

Described in this way, PISA recognises that the study of *space and shape* is open-ended, dynamic and fundamental to *mathematical literacy*. The *TWISTED BUILDING* unit is an example of a *space and shape* question that begins with the context of a geometric structure (a building), provides a more familiar mathematical representation of part of the situation, and calls on students to interpret the context, and to apply some mathematical knowledge to answer two questions examining spatial relationships from different perspectives (see Annex A1). Table 2.2 lists all of the released PISA *space and shape* questions that were used in the main PISA 2003 assessment and where the reader can find these in Chapter 3.

Table 2.2
Examples of *space and shape* questions

Question		Where to find question in Chapter 3
STAIRCASE	Question 1	Examples of easy questions section
CUBES	Question 1	
NUMBER CUBES	Question 2	Examples of questions of moderate difficulty section
CARPENTER	Question 1	Examples of difficult questions section

Quantity

PISA recognises the importance of quantitative literacy. In PISA, the overarching idea of *quantity* includes: meaning of operations, feel for magnitude of numbers, smart computations, mental arithmetic, estimations. Given the fundamental role of quantitative reasoning in applications of mathematics, as well as ubiquitous presence of numbers in our lives, it is not surprising that number concepts and skills form the core of school mathematics. In the earliest grade, mathematics teachers start children on a mathematical path designed to develop computational procedures of arithmetic together with the corresponding conceptual understanding that is required to solve quantitative problems and make informed decisions.

Quantitative literacy requires an ability to interpret numbers used to describe random as well as deterministic phenomena, to reason with complex sets of interrelated variables, and to devise and critically interpret methods for quantifying phenomena where no standard model exists.

Quantity requires an understanding of numeric phenomena, quantitative relationships and patterns.

Quantitatively literate students need a flexible ability to *(a)* identify critical relations in novel situations, *(b)* express those relations in effective symbolic form, *(c)* use computing tools to process information, and *(d)* interpret the results of these calculations (Fey, 1990).

PISA also aims to assess whether students can demonstrate creative quantitative reasoning. Creativity, coupled with conceptual understanding, is often ignored across the school curriculum. Students may have little experience in recognising identical problems presented in different formats or in identifying seemingly different problems that can be solved using the same mathematical tools. For example, in PISA quantitatively literate students would be able to recognise that the following three problems can all be solved using the concept of ratio:

- *Tonight you're giving a party. You want about a hundred cans of Coke. How many six-packs are you going to buy?*

- *A hang glider with glide ratio of 1 to 23 starts from a sheer cliff at a height of 123 meters. The pilot is aiming for a spot at a distance of 1 234 meters. Will she reach that spot?*

- *A school wants to rent minivans (with 8 seats each) to transport 78 students to a school camp. How many vans will the school need?*

Table 2.3 lists all of the released PISA *quantity* questions that were used in the main PISA 2003 assessment and where the reader can find these in Chapter 3. For example, the EXCHANGE RATE unit includes three questions with a context of travel and international exchange rates that call on students to demonstrate interpretation and quantitative reasoning skills.

Table 2.3
Examples of *quantity* questions

Question		Where to find question in Chapter 3
EXCHANGE RATE	Question 1	Examples of easy questions section
EXCHANGE RATE	Question 2	
SKATEBOARD	Question 1	
STEP PATTERN	Question 1	
BOOKSHELVES	Question 1	
SKATEBOARD	Question 3	Examples of questions of moderate difficulty section
CHOICES	Question 1	
SKATEBOARD	Question 2	
EXCHANGE RATE	Question 3	

Uncertainty

Uncertainty involves probabilistic and statistical phenomena as well as relationships that become increasingly relevant in the information society.

In PISA the overarching idea of *uncertainty* is used to suggest two related topics: statistics and probability. Both of these are phenomena that are the subject of mathematical study. Recent moves have occurred in many countries towards increasing the coverage of statistics and probability within school curricula, particularly in recognition of the increasing importance of data in modern life. However it is particularly easy for a desire to increase the focus on data analysis to lead to a view of probability and statistics as a collection of specific and largely unrelated skills. Following the definition of the well-known statistics educator David S. Moore (1990), PISA *uncertainty* recognises the importance for students to: i) view data as *numbers in a context*; ii) develop an understanding of random events, the term he uses to label phenomena having uncertain individual outcomes but a regular pattern of outcomes in many repetitions.

Studies of human reasoning have shown that a student's intuition concerning randomness and chance profoundly contradicts the laws of probability. In part, this is due to students' limited contact with randomness. The study of data offers a natural setting for such an experience.

Randomness is a concept that is hard to deal with: children who begin their education with spelling and multiplication expect the world to be deterministic. They learn quickly to expect one answer to be right and others to be wrong, at least when the answers take numerical form. Probability is unexpected and uncomfortable, as Arthur Nielsen from the famous market research firm noted:

> [Business people] accept numbers as representing Truth .... They do not see a number as a kind of shorthand for a range that describes our actual knowledge of the underlying condition. ... I once decided that we would draw all charts to show a probable range around the number reported; for example, sales are either up three per cent, or down three per cent or somewhere in between. This turned out to be one of my dumber ideas. Our clients just couldn't work with this type of uncertainty (Nielsen 1986, p. 8).

Statistical thinking involves reasoning from empirical data that are non-deterministic in nature, and should therefore be part of the mental equipment of every intelligent citizen. According to Moore (1990, p. 135) the core elements of statistical thinking involve the omnipresence of variation in processes and the need for data about processes to understand them. It also involves the need to take account of potential sources of variation when planning data collection or production, quantification of variation and explanation of variation.

Data analysis might help the learning of basic mathematics. The essence of data analysis is to "let the data speak" by looking for patterns in data, so that inferences can then be made about the underlying reality.

Table 2.4 lists all the released *uncertainty* questions that were used in the main PISA 2003 assessment and where the reader can find these in Chapter 3. For example, SUPPORT FOR THE PRESIDENT – QUESTION 1 exemplifies the statistics aspect of the *uncertainty* overarching idea. The stimulus for this question consists of information from opinion polls about a forthcoming election, conducted under varying conditions by four newspapers. Students were asked to reflect on the conditions under which the polls were conducted and to apply their understanding of such fundamental statistical concepts as randomness, and sampling procedures, and to tie these to their "common sense" ideas about polling procedures, to decide and explain which of the polls is likely to provide the best prediction.

Table 2.4
Examples of *uncertainty* questions

Question		Where to find question in Chapter 3
EXPORTS	*Question 1*	Examples of easy questions section
COLOURED CANDIES	*Question 1*	
LITTER	*Question 1*	
SCIENCE TESTS	*Question 1*	Examples of questions of moderate difficulty section
EARTHQUAKE	*Question 1*	
EXPORTS	*Question 2*	
SUPPORT FOR PRESIDENT	*Question 1*	
TEST SCORES	*Question 1*	Examples of difficult questions section
FORECAST OF RAIN	*Question 1*	
ROBBERIES	*Question 1*	

OVERARCHING IDEAS AND TRADITIONAL TOPICS

The comparison across PISA countries of student performance within each of the overarching ideas is described in the OECD report *Learning for Tomorrow's World: First Results from PISA 2003* (OECD, 2004a). This report also describes in great detail what students can typically do in these four areas of mathematics.

For certain topics and groups of countries, PISA mathematics questions are reclassified into five traditional curriculum topics: Number, Algebra, Measurement, Geometry and Data.

In Chapter 4 of the present report, the PISA items are reclassified to more traditional curriculum topics (used in the TIMSS survey). An analysis then is performed for specific mathematics topics and for specific groups of countries. In this study, the PISA mathematics items have been classified under the following five general mathematics curriculum topics: Number, Algebra, Measurement, Geometry and Data. These five curriculum topics are typically included in national curriculum documents in many countries. Table 2.5 shows the cross-classification of the 85 PISA 2003 main survey items according to the PISA and TIMSS (Grade 8) content classifications (OECD, 2003; Mullis, Martin, Gonzalez, and Chrostowski, 2004).

Table 2.5
Cross-tabulation of PISA items by PISA and traditional topics classifications

PISA Overarching ideas	Algebra	Data	Geometry	Measurement	Number	Total
Change and relationships	7	10	0	2	3	22
Quantity	0	0	0	0	23	23
Space and shape	0	1	12	6	1	20
Uncertainty	0	15	0	0	5	20
Total	7	26	12	8	32	85

However it is important to view such a cross-classification with some caution, and to keep in mind that there is not a strict correspondence between the phenomenological categories used in the PISA framework to define mathematical content, and the traditional mathematics topics listed here and used in TIMSS. Such a list of topics (as well as others not listed here) has typically been used as a way to organise mathematical knowledge for the purposes of designing and delivering a school syllabus and for assessing the mastery of specific knowledge. The much broader PISA categories arise from the way mathematical phenomena appear in the real world – typically unaccompanied by any clues as to which pieces of mathematical knowledge might be relevant, and where a variety of different kinds of mathematical approaches might be possible and valid.

CONTEXT – SETTING THE MATHEMATICAL PROBLEM TO BE SOLVED

The context in which a mathematics problem is situated plays an important role in real world problem solving and *mathematical literacy*. The role and relevance of context is often underestimated and even ignored in school mathematics. PISA recognises the importance of context, and gives it a major role in the assessment of *mathematical literacy*.

Most importantly, PISA recognises the need to include a variety of contexts in the assessment, as well as allowing for a range of roles for the contexts. The variety is needed in such a large international assessment to minimise the chance of featuring issues and phenomena that are too culturally specific, or too unbalanced in relation to particular cultures.

Variety of contexts

A wide range of contexts is encountered by citizens and it seems prudent to make use of the full range in constructing assessment tasks and in developing teaching and learning materials. The aspect of context needs further study, as results so far are inconclusive in the sense that one cannot say which contexts are more attractive for students or better suited for assessment or learning tasks. A common belief suggests that less able students "prefer" contexts closer to their immediate environment because they can engage with the context more readily. This can lead to items such as:

- *An ice cream seller has computed that for each 10 ice creams he sells, they will on average be the following kinds: 2 cups, 3 cones and 5 sticks. He orders 500 ice creams for the football game. How many of each kind should he order?*

- *Marge is lighter than Alice. Anny is lighter than Alice. Who is lighter: Anny or Marge?*

- *A pack of papers containing 500 sheets is 5 cm thick. How thick is one sheet of paper?*

At the primary level it is common to see this kind of context that is "close to the student" and taken from his or her "daily life". As students progress to the upper grades, the role of context often decreases and when it does occur, it is often as less familiar context drawn from the sciences or another discipline studied in the school curriculum. The exact role of context and its impact on student performance in assessment settings is not known. The use of items within context raises questions about differential student opportunities to learn the context interpretation skills. At the same time, the use of context situates the student assessment as close as possible to the real world contexts in which the student will be expected to make use of the mathematical content and modelling processes. Finding appropriate contexts and assuring they are bias free is a major issue related to context based assessment.

At the secondary level, an assumption that the context needs to be very close to the student does not necessarily hold. There are at least two relevant issues. First, it is necessary to recognise that there are more and more new "real-worlds" for students as their awareness and understanding grows – including the scientific and political worlds. But there also seems to be a tendency for this development to be postponed somewhat for lower-ability students. Second, it is necessary to recognise the role of context in assessments. Mathematics forms part of the real world, so students are bound to encounter mathematics to some degree and this is recognised in a small number of PISA items that have purely mathematical contexts (Linn, Baker, & Dunbar, 1991).

PISA mathematics tasks are set in a range of contexts…

… although there is debate as to whether or not contexts need to be close to the student.

The four different contexts relate to…

PISA contexts

PISA mathematics questions are set in four different contexts:

- *Personal*
- *Educational and occupational*
- *Public*
- *Scientific* (including intra-mathematical)

… day-to-day activities…

The unit SKATEBOARD contains three questions, Q1, Q2 and Q3, classified in the *personal* context. The stimulus provides information about the cost of skateboard components, and the questions ask students to perform various calculations to explore costs and options related to constructing a skateboard from those components. It is assumed that such a context would be of immediate and direct personal relevance to many 15-year-olds.

… school and work situations…

The *educational and occupational* contexts include problem situations that students might confront while at school, including those rather artificial problems designed specifically for teaching or practice purposes, or problems that would be met in a work situation. STEP PATTERN Q1 is in the former category – it is a simple problem of number patterns that could typically be used to teach ideas about mathematical sequences. BOOKSHELVES Q1 could be regarded as an example of the latter category – the stimulus refers to the components needed by a carpenter to construct a set of bookshelves.

… the wider community…

Public contexts are those situations experienced in one's day-to-day interactions with the outside world. An example is ROBBERIES Q1 which presents an item from a newspaper, and asks students to make a judgment about claims made in the article.

… and scientific or explicitly mathematical problems …

Examples of items presented in a *scientific* context can be found in the unit titled DECREASING CO_2 LEVELS. This unit was used at the field trial stage but was not included in the main survey test instrument. The stimulus for this unit presents scientific data on the level of carbon dioxide emissions for several countries, and the items ask students to interpret and make use of the data presented.

Mathematical relevance of context

Contexts can be present just to make the problem look like a real-world problem (fake context, camouflage context, "zero-order" context). PISA attempts to stay away from such uses if possible in its assessment items, however some such problems have been used. An example is CARPENTER Q1. The context for the problem is a set of shapes being considered by a carpenter as possible borders around a hypothetical garden bed. Nothing about the carpenter or the garden are needed to understand or solve the problem, they merely provide the camouflage for a geometry problem.

The real "first-order" use of context is when the context is relevant and needed for solving the problem and judging the answer. An example of this use of context is in the questions Q1 and Q2 of the unit EXCHANGE RATE, where the context of two different currencies and the conversions between them are needed in order to understand and solve the problems and to evaluate the solutions. Another example is the problem of WATER TANK Q1. Here a water tank that comprises a conical part and a cylindrical part is presented, along with five optional graphs that could represent a mathematisation of the rate of change of water height in the tank over time as the tank is filled. Students must carefully check which of the given graphs fits the context.

PISA strives to present student with only relevant contexts…

Second order use of context appears when one really needs to move backwards and forwards between the mathematical problem and its context in order to solve the problem or to reflect on the answer within the context to judge the correctness of the answer. This process is referred to as "mathematisation", and is discussed more extensively in Chapter 6. Thus, the distinction between first- and second-order uses of context lies in the role of the mathematisation process. In the first order, PISA has already pre-mathematised the problem, whereas in the second order much emphasis is placed on the mathematising process. An example of this second order use of context can be seen in TWISTED BUILDING Q1. The context of this question is a photograph of a computer model of a building, and students must impose their own mathematical structures on the situation in order to estimate the height of the building. Another example is ROCK CONCERT Q1 in which students are presented with the dimensions of a hypothetical football pitch, and have to model the space occupied by a person in order to estimate the number of football fans that could be accommodated.

… especially contexts that require students to shift between the mathematical problem and its context.

The highest levels of *mathematical literacy* involve the ability to effectively handle such second-order contexts. This is the essence of real-world problem solving. From a mathematical instruction perspective, it is essential that students are exposed to activities that involve the purposeful interpretation of contexts in order to produce a relevant mathematical representation of the underlying problem, and that require reference to the context in order to produce a solution that addresses the problem.

Having dealt with content (mathematics literacy) and contexts (situations), the analysis turns to the third 'C': the competencies.

THE COMPETENCIES

An individual who is to make effective use of his or her mathematical knowledge within a variety of contexts needs to possess a number of mathematical competencies. Together, these competencies provide a comprehensive foundation for the proficiency scales that are described further in this chapter. To identify and examine these competencies, PISA has decided to make use of eight characteristic mathematical competencies that are relevant and meaningful across all education levels.

The cognitive activities needed to solve real-world problems are divided into eight characteristic mathematical competencies.

As can be seen immediately, there is substantial overlap among these competencies. This results from the ways in which these competencies are interrelated in the application of mathematics in solving a problem. It is usually necessary to draw on several of the competencies at once in such situations.

Mathematical thinking and reasoning

A fundamental mathematical competency is the capacity to think and reason mathematically. This involves asking probing and exploratory questions about what is possible, what could happen under certain conditions, how one might go about investigating a certain situation, and analysing logically the connections among problem elements.

Mathematical argumentation

Competency related to formal and logical argument and to justification and proof is also central to *mathematical literacy*. Such competence includes the ability to follow chains of reasoning and argument and to create such chains in analysing a process mathematically. At other times, this competence emerges in explaining, justifying or proving a result.

Modelling

Another competency associated with *mathematical literacy* is modelling. It is critical to *mathematical literacy* since it underpins the capacity to move comfortably between the real world in which problems are met and solutions are evaluated, and the mathematical world where problems are analysed and solved. The modelling process includes the capacity to structure the situation to be modelled, to translate from the real world to the mathematical world, to work with the model within the mathematical domain, to test and validate models used, to reflect critically on the model and its results especially in relation to the real world situation giving rise to the modelling activity, to communicate about the model, its results, and any limitations of such results, and to monitor and control the whole modelling process.

Problem posing and solving

An important step in solving problems is the capacity to define and clarify the problem to be solved. A mathematically literate person will have competence in working with problems in such a way that facilitates formulating clear problems from a relatively unstructured and ill-defined problem-situation and then carrying out sustained thought and analysis to bring relevant mathematical knowledge to bear on the transformed problem. This might involve recognising similarities with previously solved problems, or using insight to see where existing knowledge and skills can be applied, or creative linking of knowledge and information to produce a novel response to the situation.

Representation

A very basic competency that is critically important to *mathematical literacy* is the capacity to successfully use and manipulate a variety of different kinds of representations of mathematical objects and situations. This may include such representations as graphs, tables, charts, photographs, diagrams and text, as well as, algebraic and other symbolic mathematical representations. Central to this competence is the ability to understand and make use of interrelationships among these different representations.

Symbols and formalism

A defining competency of *mathematical literacy* is the capacity to understand and use mathematical symbolic language. This includes decoding symbolic language and understanding its connection to natural language. More generally, this competency also relates to the ability to handle and work with statements containing symbols and formulas, as well as the technical and procedural mathematical skills associated with a wide variety of formal mathematical processes.

Communication

Mathematical literacy is also about competence in communication – understanding the written, oral, or graphical communications of others about mathematical matters and the ability to express one's own mathematical views in a variety of ways.

Aids and tools

The competency associated with knowing about and being able to make use of various aids and tools, including information technology tools, is another important part of *mathematical literacy*, particularly where mathematical instruction is concerned. Students need to recognise when different tools might be useful, to be able to make appropriate use of those tools, and to recognise the limitations of those tools.

It should be recognised that any effort to assess individual competencies is likely to result in artificial tasks and unnecessary and undesirable compartmentalisation of the *mathematical literacy* domain. In order to productively describe and report student's capabilities, as well as their strengths and weaknesses from an international perspective, some structure is needed.

COMPETENCY CLUSTERS

PISA competencies are classified into three clusters...

When doing real mathematics, it is necessary to draw simultaneously upon many of those competencies. In order to operationalise these mathematical competencies, PISA groups the underlying skills into three competency clusters:

1. Reproduction

2. Connections

3. Reflection

Reproduction cluster

... those involving familiar mathematical processes and computations ...

Questions in the *reproduction* competency cluster require students to demonstrate that they can deal with knowledge of facts, recognise equivalents, recall mathematical objects and properties, perform routine procedures, apply standard algorithms and apply technical skills. Students also need to deal and operate with statements and expressions that contain symbols and formulas in "standard" form. Assessment items from the *reproduction* cluster are often in multiple-choice, fill-in-the-blank, matching, or (restricted) open-ended formats.

Table 2.6 lists all of the released questions in the *reproduction* competency cluster that were used in the PISA 2003 main assessment. Each of these is presented in full in Chapter 3 along with a description of the particular competencies that students need to draw upon to successfully solve these mathematical problems.

Table 2.6
Examples of questions in the *reproduction* competency cluster

Question	Where to find question in Chapter 3
EXCHANGE RATE	Question 1
STAIRCASE	Question 1
EXPORTS	Question 1
EXCHANGE RATE	Question 2
THE BEST CAR	Question 1	Examples of easy questions section
GROWING UP	Question 1
GROWING UP	Question 2
CUBES	Question 1
SKATEBOARD	Question 1
STEP PATTERN	Question 1
COLOURED CANDIES	Question 1	Examples of questions of moderate difficulty section
SCIENCE TESTS	Question 1
SKATEBOARD	Question 2
WALKING	Question 1	Examples of difficult questions section

Connections cluster

Questions in the *connections* competency cluster require students to demonstrate that they can make linkages between the different strands and domains within mathematics and integrate information in order to solve simple problems in which students have choices of strategies or choices in their use of mathematical tools. Questions included in the *connections* competency cluster are non-routine, but they require relatively minor amounts of translation between the problem context and the mathematical world. In solving these problems students need to handle different forms of representation according to situation and purpose, and to be able to distinguish and relate different statements such as definitions, claims, examples, conditioned assertions and proof.

… those involving a degree of interpretation and linkages…

Students also need to show good understanding of mathematical language, including the decoding and interpreting of symbolic and formal language and understanding its relations to every-day language. Questions in the *connections* competency cluster are often placed within a *personal*, *public* or *educational and occupational* context and engage students in mathematical decision-making.

Table 2.7 lists the all of the released questions in the *connections* competency cluster that were used in the PISA 2003 main assessment. Each of these is presented in full in Chapter 3 along with a description of the particular competencies that students need to draw upon to successfully solve these mathematical problems.

Table 2.7
Examples of questions in the *connections* competency cluster

Question	Where to find question in Chapter 3	
BOOKSHELVES	Question 1	
NUMBER CUBES	Question 2	
INTERNET RELAY CHAT	Question 1	
SKATEBOARD	Question 3	Examples of questions of moderate difficulty section
CHOICES	Question 1	
EXPORTS	Question 2	
GROWING UP	Question 3	
SUPPORT FOR PRESIDENT	Question 1	
TEST SCORES	Question 1	
FORECAST OF RAIN	Question 1	Examples of difficult questions section
ROBBERIES	Question 1	
WALKING	Question 3	
CARPENTER	Question 1	

Reflection cluster

Questions in the *reflection* competency cluster typically present students with a relatively unstructured situation, and ask them to recognise and extract the mathematics embedded in the situation and to identify and apply the mathematics needed to solve the problem. Students must analyse, interpret, develop their own models and strategies, and make mathematical arguments including proofs and generalisations. These competencies include a critical component that involves analysis of the model and reflection on the process.

… and those involving deeper insights and reflection.

Questions in the *reflection* competency cluster also require students to demonstrate that they can communicate effectively in different ways (*e.g.* giving explanations and arguments in written form, or perhaps using visualisations). Communication is meant to be a two-way process: students also need to be able to understand communications produced by others that have a mathematical component.

Table 2.8 lists all of the released questions in the *reflection* competency cluster that were used in the PISA 2003 main assessment. Each of these is presented in full in Chapter 3 along with a description of the particular competencies that students need to draw upon to successfully solve these mathematical problems.

Table 2.8
Examples of questions in the *reflection* competency cluster

Question		Where to find question in Chapter 3
LITTER	*Question 1*	
EARTHQUAKE	*Question 1*	Examples of questions of moderate difficulty section
EXCHANGE RATE	*Question 3*	
INTERNET RELAY CHAT	*Question 2*	Examples of difficult questions section
THE BEST CAR	*Question 2*	

CONCLUSION

Mathematics in PISA 2003 focused on 15-year-olds' capabilities to use their mathematical knowledge in solving mathematical situations presented in a variety of settings. This focus was on their functional use of knowledge in solving real-life problems, rather than on ascertaining to what degree they had mastered their studies of formal mathematics or the degree to which they were facile with particular facts or procedures. This focus on *mathematical literacy* is examined through the lenses of student achievement related to situations calling for knowledge of *change and relationships*, *space and shape*, *quantity*, and *uncertainty*. These more global overarching ideas are supplemented by analyses taking the form of traditional topics such as algebra and functions, geometry and measurement, etc.

As the PISA 2003 study did not have teacher questionnaires providing information on the actual implementation of curriculum or teaching processes in the students' classrooms, PISA mathematics examines the student performance through the added lenses of context and competencies. Items were classified by context according to whether the situations contained were presented as dealing with their *personal* life, with a possible *educational* or *occupational* task, with a *public* use of mathematics task, or with an application of mathematics in a *scientific* or, even, mathematical setting. At the same time, items were classified by the demands they placed on students' cognitive processing capabilities. These demands were identified by the competencies discussed and their amalgamation into the clusters of *reproduction*, *connections*, and *reflection*.

The examination of the content through the filters of content, context, and competencies provides a filtering that helps understand the mathematical capabilities that students have developed in their first 15 years of life. Capabilities which are based in part on formal educational experiences, but which, in many cases, result from their direct experiences with solving problems that arise in daily life and decision making. This focus on *mathematical literacy* using the PISA definition strengthened by the three competencies presents a unique view of what students know and are able to do when confronted with situations to which mathematics knowledge and skills are applicable. The following chapters will detail the findings of the applications of this framework to the PISA 2003 mathematics results.

A Question of Difficulty:
Questions from PISA 2003

> *This chapter illustrates the PISA 2003 assessment with released assessment items and links to different levels of mathematical literacy proficiency. Actual assessment items can be found in this chapter, along with a discussion of students' performance on each of them.*

INTRODUCTION

This chapter presents a number of characteristics of the PISA questions in relation to different levels of proficiency in mathematics. The characteristics discussed include the proficiency descriptions used to report the different levels of performance of students in the PISA mathematics assessment and the related issue of how difficult the question is, the type of response required by students (*e.g.* select a given response or write a short answer), and the role of the context that the question is set in. The complexity of the language used and other aspects of the presentation of questions will also be discussed.

DESCRIBING GROWTH IN MATHEMATICAL LITERACY: HOW DIFFICULT IS THE QUESTION AND WHERE DOES IT FIT ON THE PISA MATHEMATICS SCALE?

The questions used in the PISA mathematics assessment cover a wide range of difficulties. This is necessary in order to obtain valid and reliable ability estimates for the range of students sampled in different countries. The difficulty of the questions used can be illustrated by reference to the PISA mathematics scale that was developed to quantify performance in different countries (OECD, 2004a). This chapter discusses factors that contribute to the difficulty of questions in PISA mathematics.

PISA mathematics questions cover a wide range of difficulties in a wide range of formats.

The PISA mathematics questions take a variety of formats, and while Chapter 5 analyses more extensively the relationship between the type of question and how difficult the question is, the basic types of PISA mathematics questions are briefly introduced here. In the 2003 assessment, all mathematics questions broadly either required students to construct a response or to select a response. In the case of the latter, these could be either simple multiple-choice questions, requiring the students to select one answer from a number of optional responses, or complex multiple-choice questions, presenting students with a small number of statements and requiring students to select from given optional responses for each statement (such as "true" or "false"). In the case of questions where students need to construct a response, this could be either an extended response (*e.g.* extensive writing, showing a calculation, or providing an explanation or justification of the solution given) or a short answer (*e.g.* a single numeric answer, or a single word or short phrase; and sometimes a slightly more extended short response). Much of the discussion around reform in mathematics education involves questions presented in context and requiring communication as part of the response (de Lange, 2007). The analyses of item difficulty in this chapter and, later, in Chapters 4 and 5, focuses on how questions were presented to students and the degree to which students were able to meet the challenges posed by the items.

The methodology of the PISA assessment, including the sampling design, the design of the assessment instruments including the various types of questions, and the methods used to analyse the resulting data, leads to efficient estimates of the proportion of students in each country lying at various parts of the *mathematical literacy* scale. *Mathematical literacy* is conceived as a continuous variable, and the scale has been developed to quantify and describe this. The PISA *mathematical literacy* scale is constructed to have a mean of 500 score points and a standard deviation of 100 score points; that is, about two-thirds of the 15-year-olds across OECD countries score between 400 and 600 score points. Six proficiency levels are defined for the *mathematical literacy* scale, and the kinds of student behaviours typical in each of those proficiency levels are described. This "described proficiency scale" is central to the way in which PISA reports comparative performance in mathematics.

The difficulty of PISA mathematics questions is determined using three different approaches…

This report uses three different but related methods to quantify and refer to the difficulty of the mathematics items. First, the simplest approach involves using "percent correct" data (that is, the percentage of students in each country or internationally correctly answering a question). This form of comparison is useful when the focus is on an individual question (for example, comparing the success rate of male and female students on a particular item, or of students in different countries on a particular item), or on comparing the performance on two questions by a particular group of students.

… with simple percentages …

Second, the formal statistical analysis of PISA data is carried out using units called a "logit". A logit represents the logarithm of the ratio of the probability of a correct answer to an item to the probability of an incorrect answer [often called the log-odds ratio]. For example an item with a probability correct of 0.50 would have a logit value equal to 0 [log (0.5/0.5) = log (1) = 0]. The use of log-odds ratio transforms the infinite scale associated with the probability ratio through logarithms to a 0-1 scale estimation of the location of the difficulty of all items and the ability of all students on a single dimension. Item performance can then be placed on a single scale by their log-odds ratio. This approach is basic to item response theory and its depiction of item difficulty and other item parameters through varied parameter models using the logistic function. In particular, the use of logit scores for items places them on a linear scale allowing for arithmetic computations with the logit unit (Thissen & Wainer, 2001). This is useful in comparing the relative strengths and weaknesses of items, and students on the PISA mathematics framework within each country and is discussed extensively in Chapter 4. Third, the "logits" units for each question are transformed as described to form the PISA mathematics scale giving an associated score in PISA score points. This approach allows for each PISA mathematics question to be located along the same scale and thus shows the relative difficulty of each question.

… logistic models …

… and the statistically calculated PISA mathematics scale.

THE PISA SCALE AND DIFFICULTY

Figures 3.1a, 3.1b and 3.1c illustrate the placement of items on the PISA mathematics examination in terms of their relation to the PISA scale's scores. Released items and student performance on them are illustrated in the following sections in an explanation of student performance associated with various intervals on the PISA mathematics scale.

PISA releases some questions after the assessment to help illustrate the kind of mathematics problems that students have to solve. Thirty-one of the mathematics questions used in the PISA 2003 assessment were publicly released and

A Question of Difficulty: Questions from PISA 2003

Figure 3.1a ■ PISA mathematics proficiency Levels 1 and 2: Competencies students typically show and publicly released questions

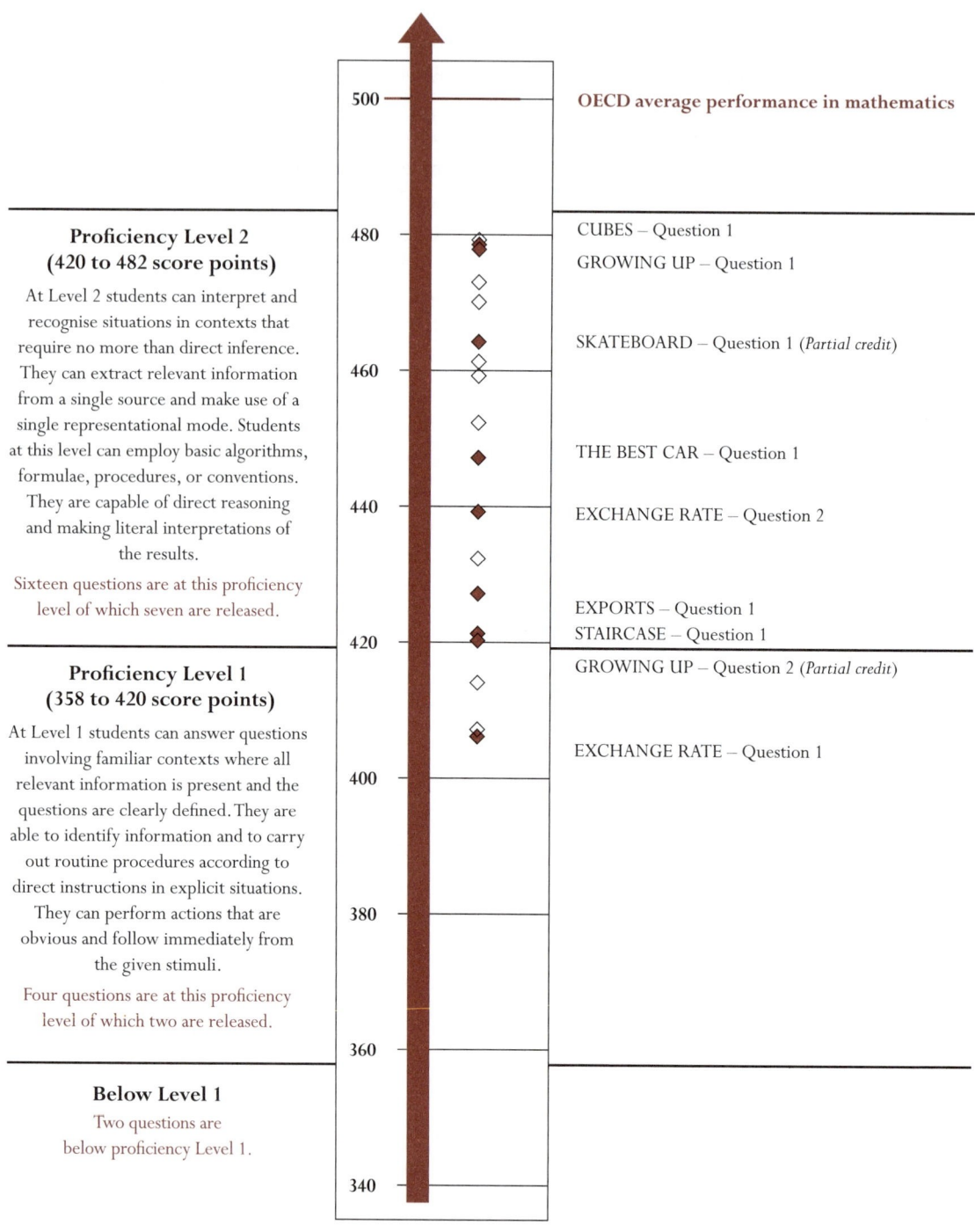

OECD average performance in mathematics — 500

Proficiency Level 2 (420 to 482 score points)

At Level 2 students can interpret and recognise situations in contexts that require no more than direct inference. They can extract relevant information from a single source and make use of a single representational mode. Students at this level can employ basic algorithms, formulae, procedures, or conventions. They are capable of direct reasoning and making literal interpretations of the results.

Sixteen questions are at this proficiency level of which seven are released.

Score	Question
480	CUBES – Question 1
	GROWING UP – Question 1
	SKATEBOARD – Question 1 (*Partial credit*)
460	
	THE BEST CAR – Question 1
440	EXCHANGE RATE – Question 2
	EXPORTS – Question 1
420	STAIRCASE – Question 1

Proficiency Level 1 (358 to 420 score points)

At Level 1 students can answer questions involving familiar contexts where all relevant information is present and the questions are clearly defined. They are able to identify information and to carry out routine procedures according to direct instructions in explicit situations. They can perform actions that are obvious and follow immediately from the given stimuli.

Four questions are at this proficiency level of which two are released.

GROWING UP – Question 2 (*Partial credit*)

EXCHANGE RATE – Question 1

400
380
360

Below Level 1

Two questions are below proficiency Level 1.

340

Twenty questions with scores at these levels.

42 Learning Mathematics for Life: A Perspective from PISA – © OECD 2009

Figure 3.1b ■ **PISA mathematics proficiency Levels 3 and 4:**
Competencies students typically show and publicly released questions

Proficiency Level 4
(544 to 607 score points)

At Level 4 students can work effectively with explicit models for complex concrete situations that may involve constraints or call for making assumptions. They can select and integrate different representations, including symbolic ones, linking them directly to aspects of real-world situations. Students at this level can utilise well-developed skills and reason flexibly, with some insight, in these contexts. They can construct and communicate explanations and arguments based on their interpretations, arguments, and actions.

Twenty-six questions are at this level of which twelve are released.

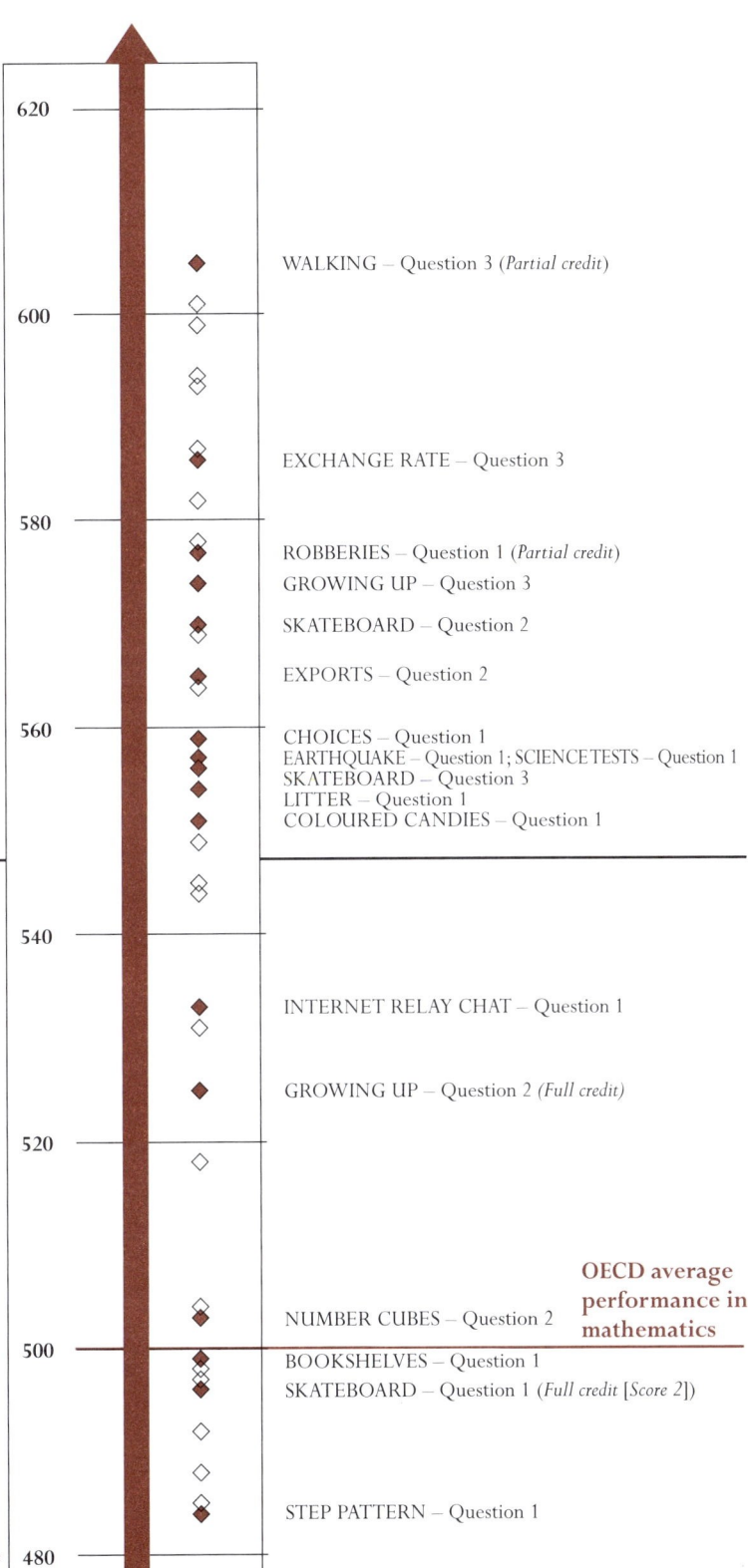

Proficiency Level 3
(482 to 544 score points)

At Level 3 students can execute clearly described procedures, including those that require sequential decisions. They can select and apply simple problem-solving strategies. Students at this level can interpret and use representations based on different information sources and reason directly from them. They can develop short communications reporting their interpretations, results and reasoning

Seventeen questions are at this level of which six are released.

Forty-three questions with scores at these levels, of which eighteen are released.

Score	Question
~605	WALKING – Question 3 *(Partial credit)*
~585	EXCHANGE RATE – Question 3
~577	ROBBERIES – Question 1 *(Partial credit)*
~574	GROWING UP – Question 3
~570	SKATEBOARD – Question 2
~565	EXPORTS – Question 2
~559	CHOICES – Question 1
~557	EARTHQUAKE – Question 1; SCIENCE TESTS – Question 1
~554	SKATEBOARD – Question 3
~552	LITTER – Question 1
~551	COLOURED CANDIES – Question 1
~533	INTERNET RELAY CHAT – Question 1
~525	GROWING UP – Question 2 *(Full credit)*
~503	NUMBER CUBES – Question 2
~499	BOOKSHELVES – Question 1
~496	SKATEBOARD – Question 1 *(Full credit [Score 2])*
~484	STEP PATTERN – Question 1

OECD average performance in mathematics (line at 500)

A Question of Difficulty: Questions from PISA 2003

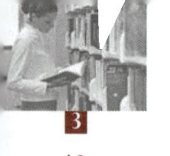

Figure 3.1c ■ **PISA mathematics proficiency Levels 5 and 6: Competencies students typically show and publicly released questions**

Proficiency Level 6
(669 score points and above)

Students can conceptualise, generalise, and utilise information based on their investigations and modelling of complex problem situations. They can link different information sources and representations and flexibly translate among them. Students at this level are capable of advanced mathematical thinking and reasoning. These students can apply this insight and understandings along with a mastery of symbolic and formal mathematical operations and relationships to develop new approaches and strategies for attacking novel situations. Students at this level can formulate and precisely communicate their actions and reflections regarding their findings, interpretations, arguments, and the appropriateness of these to the original situations.

Ten questions are at this level of which three are released.

Proficiency Level 5
(607 to 669 score points)

At Level 5 students can develop and work with models for complex situations, identifying constraints and specifying assumptions. They can select, compare, and evaluate appropriate problem solving strategies for dealing with complex problems related to these models. Students at this level can work strategically using broad, well-developed thinking and reasoning skills, appropriate linked representations, symbolic and formal characterisations, and insight pertaining to these situations. They can reflect on their actions and formulate and communicate their interpretations and reasoning.

Seventeen questions are at this level of which six are released.

Twenty-seven questions with scores at these levels.

- 801 score points
- WALKING – Question 3 (*Full credit*)
- ROBBERIES – Question 1 (*Full credit*)
- CARPENTER – Question 1 (*Full credit*)
- WALKING – Question 3 (*Partial credit*)
- THE BEST CAR – Question 2 (*Full credit*)
- INTERNET RELAY CHAT – Question 2 (*Full credit*)
- TEST SCORES – Question 1 (*Full credit*)
- SUPPORT FOR THE PRESIDENT – Question 1 (*Full credit*)
- WALKING – Question 1 (*Full credit*)

Figures 3.1a, 3.1b and 3.1c show where each of these questions is located on the PISA *mathematical literacy* scale. It is useful to remember that the OECD average performance in PISA 2003 mathematics is 500 score points. Most of the questions in Figures 3.1a, 3.1b and 3.1c involve simple scoring, where credit is awarded only if the answer is correct and a 0 is awarded otherwise. However, five of these questions involve the use of up to three different scoring categories. For these questions, the term "full credit" is used to describe a fully correct answer, and one or more "partial credit" categories exist for answers that are only partially correct, for example the student may have only solved the first step of the problem at hand or have shown all necessary working, but made a minor calculation error. As a result, for the 31 questions in Figures 3.1a, 3.1b and 3.1c result in a total of 36 different scores as shown in Figures 3.1a, 3.1b, and 3.1c. Student performance through these score levels helps illustrate the full range of PISA mathematics proficiency (Levels 1 to 6, where Level 1 is the simplest and Level 6 the hardest). Figure 3.1a shows the summary descriptions of what students can typically do at PISA mathematics proficiency Levels 1 and 2 where the easiest questions in the mathematics assessment are located. The PISA score points for all questions included in Levels 1 and 2 are below the OECD average performance of 500 score points. They range from 358 to 482 score points.

Student performance is measured on a scale with an average score of 500. Students are grouped in six levels of proficiency, plus a group below Level 1.

The remainder of the chapter presents the 31 released PISA 2003 mathematics questions to illustrate more fully the different levels of proficiency in mathematics and to analyse the characteristics related to the difficulty of the question. Questions are presented in three distinct sections: the easiest questions in PISA 2003 mathematics illustrating PISA proficiency at Levels 1 and 2 (in fact the two easiest questions in the test lie below Level 1) which are found on the PISA scale from 358 to 482 points; questions of moderate difficulty in PISA 2003 mathematics illustrating proficiency at Levels 3 and 4, which are found on the PISA scale from 482 to 607 points; and the most difficult questions in PISA 2003 mathematics illustrating proficiency at Levels 5 and 6 which are found on the PISA scale from 607 and above. In each section an introductory summary table presents the following key characteristics for all questions: the associated PISA score points on the *mathematical literacy* scale (including, where appropriate, scores for both full and partial credit); where the question fits into the three main components of the PISA mathematics framework – content area or "overarching idea", competency cluster and context; the format used for the question; the traditional mathematics topic tested most prominently in the question; and the length of question (as measured by a simple word count) to indicate the reading demand. Additional information and data on the test items and related student performance can be found at *www.pisa2003.acer.edu.au/downloads.php* at the Australian Council for Educational Research's PISA website for PISA 2003.

EXAMPLES OF THE EASIEST MATHEMATICS QUESTIONS FROM PISA 2003

In the PISA 2003 mathematics assessment, the two easiest questions lie below Level 1, and 20 questions are included in proficiency Levels 1 and 2. Nine of these 20 questions are released (coming from seven units) and these are listed in Table 3.1 along with the difficulty of each question on the PISA scale and other characteristics. Several of these questions are included in units that contain more than one question. In the case that these questions are located at different proficiency Levels (*e.g.* the unit EXPORTS contains Question 1 at Level 2 and Question 2 at Level 4) each question is presented at the section of the chapter related to where its score points appear on the PISA scale. Figures showing country level performances on many of these items are found in Annex A1, Figures A1.1 through A1.8.

Recall from Figure 3.1a that Level 1 proficiency indicates that students can answer questions involving familiar contexts where all relevant information is present and the questions are clearly defined. They are able to identify information and to carry out routine procedures according to direct instructions in explicit situations. They can perform actions that are obvious and follow immediately from the given stimuli. Level 2 students can interpret and recognise situations in contexts that require no more than direct inference. They can extract relevant information from a single source and make use of a single representational mode. Students at this level can employ basic algorithms, formulae, procedures, or conventions. They are capable of direct reasoning and making literal interpretations of the results.

Table 3.1 shows that there is a prevalence of questions in the *reproduction* competency cluster among the easiest questions in PISA 2003 mathematics. Fourteen of the 20 questions at Levels 1 and 2 are in the *reproduction* competency cluster and this is also true of the two easiest questions lying below Level 1. In general, questions in the *reproduction* competency cluster place lower-level cognitive demands on students, and are therefore easier. Nevertheless there are relatively easy questions also from the *connections* competency cluster (six of the 20 questions in Levels 1 and 2 are in this category). All four content areas are represented among the easier questions in the PISA 2003 mathematics assessment: seven questions belong to *quantity*, six to *change and relationships*, five to *space and shape* and two to *uncertainty*. However, all nine released items for Levels 1 and 2 employed the short response item format.

Each of the items listed in Figure 3.1a as appearing in Levels 1 and 2 is now examined in detail, with performance information for students from the participating countries used as a lens to understand both student work and differences between the countries. In addition to the presentation of each unit in the right-hand side of the following displays, the scoring guide with sample responses for each level is presented beneath the questions contained within each question within the unit. Additional information about the items and about technical aspects associated with the scaling of the scores can be found in the international report (OECD, 2004a) and the technical report detailing the operational aspects of the PISA 2003 study (OECD, 2005).

Table 3.1
Characteristics of the easiest released PISA 2003 mathematics questions

Item code	Question	OECD average percent correct	Question (PISA score points)	Full/Partial credit points 1	2	3	Traditional topic	Content area ("Overarching Idea")	Competency cluster	Context ("Situation")	Length of question[1]	Response format
M413Q01	EXCHANGE RATE – Question 1	80	406	406			Number	Quantity	Reproduction	Public	Medium	Short Answer
M150Q02[2]	GROWING UP – Question 2 Partial 1 Point	69	472	420	525		Data	Change and relationships	Reproduction	Scientific	Medium	Short Answer
M547Q01	STAIRCASE – Question 1	78	421	421			Number	Space and shape	Reproduction	Educational and occupational	Short	Short Answer
M438Q01	EXPORTS – Question 1	79	427	427			Data	Uncertainty	Reproduction	Public	Medium	Short Answer
M413Q02	EXCHANGE RATE – Question 2	74	439	439			Number	Quantity	Reproduction	Public	Short	Short Answer
M704Q01	THE BEST CAR – Question 1	73	447	447			Algebra	Change and relationships	Reproduction	Public	Long	Short Answer
M520Q01	SKATEBOARD – Question 1 Partial 1 Point	72	480	464	496		Number	Quantity	Reproduction	Personal	Long	Short Answer
M150Q01	GROWING UP – Question 1	67	477	477			Number	Change and relationships	Reproduction	Scientific	Short	Short Answer
M145Q01	CUBES – Question 1	68	478	478			Data	Space and shape	Reproduction	Educational and occupational	Medium	Short Answer

1. Short questions contain fewer than 50 words. Medium-length questions contain 51 to 100 words. Long questions contain more than 100 words; word count in relation to question difficulty is discussed in detail in Chapter 5.
2. Note items with partial credit are presented where that partial credit fits and the full discussion is contained later in the chapter at the level where full credit for the item is obtained. The bold type in the score column indicates the score being represented in a given problem line.

EXCHANGE RATE

Mei-Ling from Singapore was preparing to go to South Africa for 3 months as an exchange student. She needed to change some Singapore dollars (SGD) into South African rand (ZAR).

Question 1: EXCHANGE RATE

Mei-Ling found out that the exchange rate between Singapore dollars and South African rand was:

1 SGD = 4.2 ZAR

Mei-Ling changed 3000 Singapore dollars into South African rand at this exchange rate.

How much money in South African rand did Mei-Ling get?

Answer: ....

EXCHANGE RATE – Question 1 was the third easiest question of all the PISA 2003 mathematics questions. On average across OECD countries, about 80% of students solved this problem correctly.

Context: Public – currency exchange associated with international travel

Content area: Quantity – quantitative relationships with money

Competency cluster: Reproduction

The question requires students to:

- Interpret a simple and explicit mathematical relationship.
- Identify and carry out the appropriate multiplication.
- Reproduce a well-practised routine procedure.

Students were most successful on this question in the partner country Liechtenstein (95%), the partner economy Macao-China (93%), Finland (90%), and France and the partner economy Hong Kong-China (89%). Most students attempted to answer this question, with only 7% failing to respond, on average, across OECD countries.

Question 2: EXCHANGE RATE

On returning to Singapore after 3 months, Mei-Ling had 3 900 ZAR left. She changed this back to Singapore dollars, noting that the exchange rate had changed to:

1 SGD = 4.0 ZAR

How much money in Singapore dollars did Mei-Ling get?

Answer: ....

EXCHANGE RATE – Question 2 is slightly more difficult, but still among the easiest of the PISA 2003 mathematics questions. On average across OECD countries, about 74% of students were able to do this successfully. This Level 2 item had a PISA difficulty level of 439.

Context: *Public* – currency exchange associated with international travel

Content area: *Quantity* – quantitative relationships with money

Competency cluster: *Reproduction*

The question requires students to:

- Recognise the change in the context from Question 1 that results from the need to convert money in the "opposite direction".
- Carry out a division to find the required answer.

Students were most successful on this question in the partner country Liechtenstein (93%), the partner economy Macao-China (89%), Finland and the partner economy Hong Kong-China (88%), Austria (87%), and France, Switzerland and the Slovak Republic (85%). There was a 9% non-response rate across the OECD countries as a whole, while 14-17% of students in Turkey, Italy, Norway, Portugal, Greece and the partner countries Uruguay and Tunisia, failed to respond. In the partner country Brazil, 27% of students failed to respond. This compares to less than 5% of students in Finland, Canada, the Netherlands and the partner countries/economies Macao-China, Liechtenstein and Hong Kong-China.

Note that Question 3 of this unit is presented in the section *Examples of moderate to difficult questions in the PISA 2003 mathematics assessment*.

GROWING UP

Youth grows taller

In 1998 the average height of both young males and young females in the Netherlands is represented in this graph.

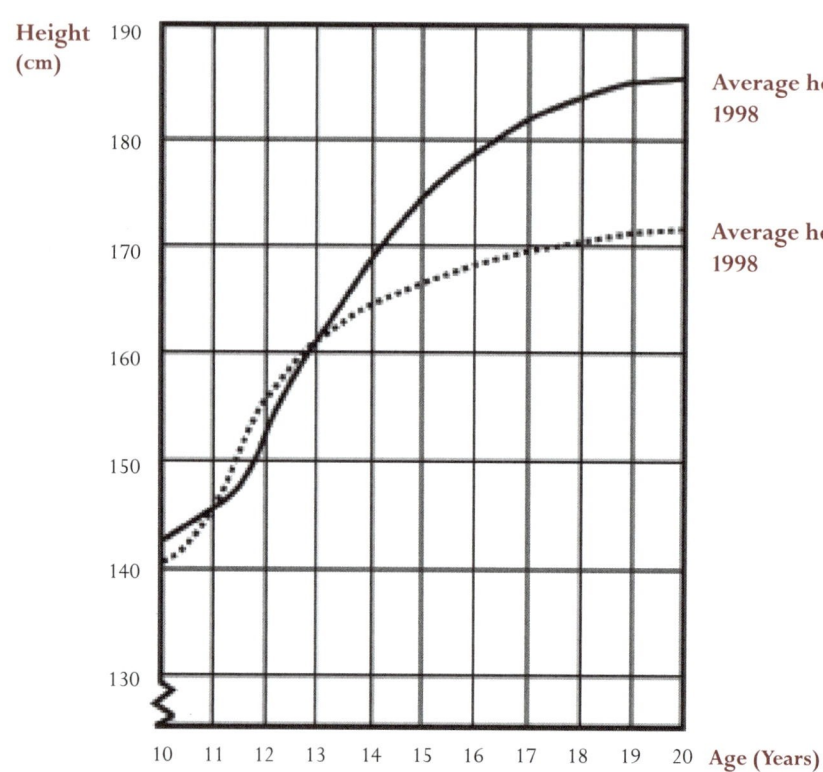

Question 1: GROWING UP

Since 1980 the average height of 20-year-old females has increased by 2.3 cm, to 170.6 cm.

What was the average height of a 20-year-old female in 1980?

Answer: .... cm

GROWING UP – Question 1 illustrates Level 2 in PISA 2003 mathematics and has a difficulty of 477 PISA score points. On average across OECD countries, 67% of students were able to do this successfully.

Context: *Scientific* – the growth curves of young males and females over a period of ten years. Science is no different from the real world in the sense that it uses graphical representation frequently, for example the graph in this question representing changes in height in relation to age.

Content area: *Change and relationships* – focus on change in height in relation to age. Basic mathematical operation of subtraction.

Competency cluster: *Reproduction* – basic thinking and reasoning involving the most basic questions (How much is the difference?); basic argumentation where the student just needs to follow a standard quantitative process. There is some added complexity in the fact that the answer can be found by ignoring the graph altogether – an example of redundant information.

The question requires students to:

- Extract the relevant information from a single source (and ignore the graph which is a redundant source).
- Make use of a single representational mode.
- Employ a basic subtraction algorithm (170.6 – 2.3).

Students were most successful on this question in Korea (82%), France (80%), Japan and the partner country the Russian Federation (78%), Sweden and Iceland (76%), the Czech Republic (75%) and the Slovak Republic (74%).

Most students attempted to answer this question – only 8% failed to do so across OECD countries and this concerned less than 1% of students in the Netherlands. However, 23% of students in Greece and 21% of students in the partner country Serbia did not respond to this question.

Question 2: GROWING UP

According to this graph, on average, during which period in their life are females taller than males of the same age?

....

GROWING UP – Question 2 illustrates two different levels of proficiency depending on whether students gave a fully or partially correct answer. Here, a partially correct answer scored at 1 point illustrates exactly the boundary between Level 1 and Level 2 with a difficulty of 420 PISA score points. A fully correct answer illustrates Level 3 with a difficulty of 525 score points. On average across OECD countries, 28% of students were only capable of achieving the partial 1 point level.

Context: *Scientific*

Content area: *Change and relationships* – focus on the relationship between age and height. The mathematical content can be described as belonging to the "data" domain: the students are asked to compare characteristics of two data sets, interpret these data sets and draw conclusions.

Competency cluster: *Reproduction* – interpret and decode reasonably familiar and standard representations of well known mathematical objects. Students need to think and reason (where do the graphs have common points?), use argumentation to explain which role these points play in finding the desired answer and communicate and explain the argumentation. However, all these competencies essentially involve reproduction of practised knowledge.

The question requires students to:

- Interpret and use a graph. Make conclusions directly from a graph. Report the results of their reasoning in a precise manner.

Students were considered to give a partially correct answer if they properly identified ages like 11 and/or 12 and/or 13 as being part of the answer, but failed to identify the continuum from 11 to 13 years. These students were able to compare the two graphs properly, but did not communicate their answer adequately or failed to show sufficient insight into the fact that the answer would be an interval. This is probably in part due to the fact that the proper procedure may not have been routine. On average across the OECD countries, 28% of students gave a partially correct answer showing that their reasoning and/or insight was well directed, but failed to come up with a full, comprehensive answer. This was the case for 43% of students in the United States, 42% of students in the Slovak Republic and the partner country Thailand, 40% of students in Poland, and between 39 and 37% of students in Italy, the Czech Republic, Sweden and the partner countries/economies the Russian Federation and Macao-China.

Seven percent of students on average across the OECD countries did not attempt to answer this question. This concerned less than 3% of students in the Netherlands, Finland, Canada and the partner economy Macao-China.

Note that the discussion of full credit for GROWING UP Q1 will be presented in the discussion of Level 3 questions in the section *Examples of moderate to difficult questions in the PISA 2003 mathematics assessment*.

STAIRCASE

Question 1: STAIRCASE

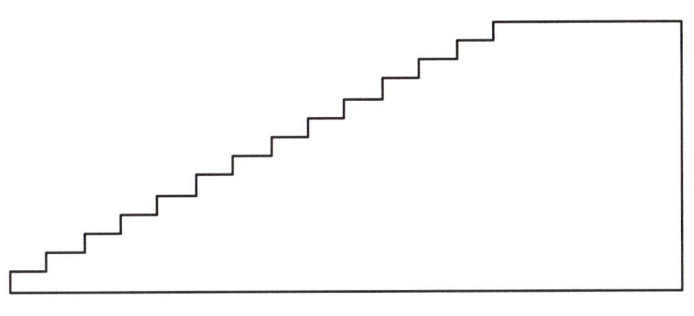

Total height 252 cm

Total depth 400 cm

The diagram above illustrates a staircase with 14 steps and a total height of 252 cm:

What is the height of each of the 14 steps?

Height: .... cm.

STAIRCASE – Question 1 illustrates Level 2 in PISA 2003 mathematics, with a difficulty of 421 PISA score points (just one point over the boundary of Level 1 and 2). On average across OECD countries, 78% of students were able to do this successfully.

Context: *Educational and occupational* – situated in a daily life context for carpenters (for example). One does not need to be a carpenter to understand the relevant information; it is clear that an informed citizen should be able to interpret and solve a problem like this that uses two different representation modes: language, including numbers, and a graphical representation. But the illustration serves a simple and non-essential function: students know what stairs look like. This item is noteworthy because it has redundant information (the depth is 400 cm) which is sometimes considered by students as confusing but a common feature in real-world problem solving.

Content area: *Space and shape* – graphical representation of a staircase, but the actual procedure to carry out is a simple division.

Competency cluster: *Reproduction* – carry out a basic operation. Students solve the problem by invoking and using standard approaches and procedures in one way only. All the required information, and even more than required, is presented in a recognisable situation.

The question requires students to:

- Extract the relevant information from a single source.

- Apply of a basic algorithm (divide 252 by 14).

In each OECD country the majority of students gave the correct answer "18", but this was especially true in the partner economy Macao-China (89%), the partner economy Hong Kong-China (87%), Switzerland (86%), Finland, the Netherlands and the partner country Liechtenstein (85%). On average across OECD countries 10% of students did not respond to this question. However in Hungary 29% of students did not respond the question, as did 25-26% of students in the partner countries Indonesia, Brazil and Thailand.

EXPORTS

The graphics below show information about exports from Zedland, a country that uses zeds as its currency.

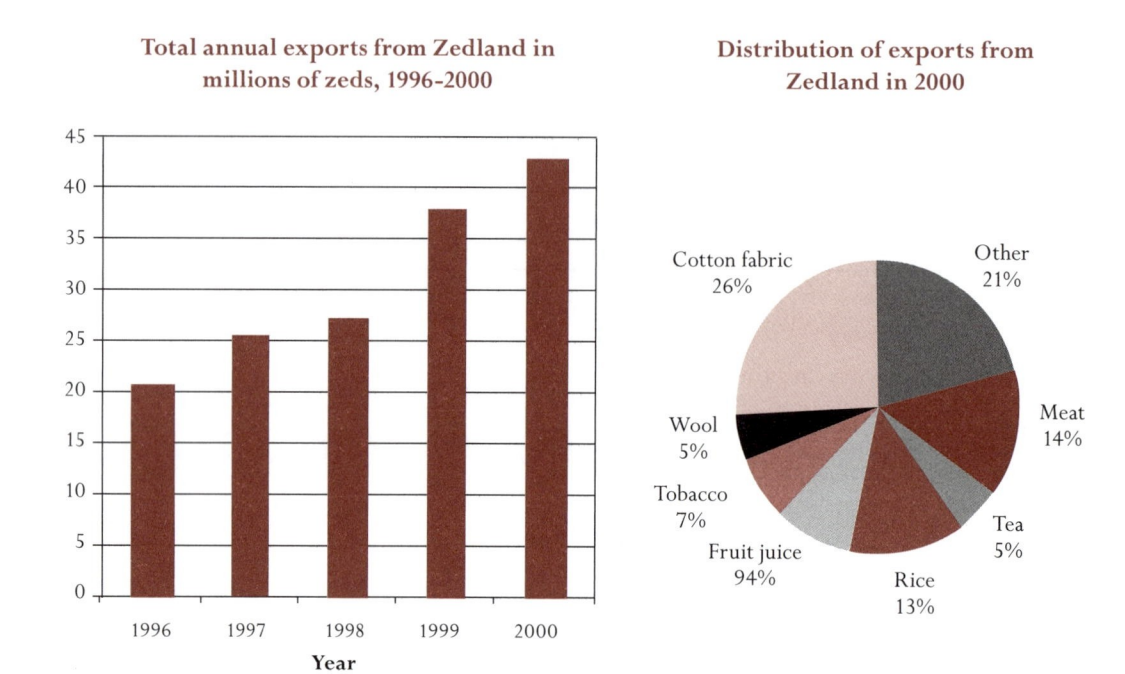

Total annual exports from Zedland in millions of zeds, 1996-2000

Distribution of exports from Zedland in 2000

- Cotton fabric 26%
- Other 21%
- Meat 14%
- Tea 5%
- Rice 13%
- Fruit juice 94%
- Tobacco 7%
- Wool 5%

Question 1: EXPORTS

What was the total value (in millions of zeds) of exports from Zedland in 1998?

Answer: ....

EXPORTS – Question 1 illustrates Level 2 in PISA 2003 mathematics, with a difficulty of 427 PISA score points. On average across OECD countries, 79% of students were able to do this successfully.

Context: *Public* – The information society in which we live relies heavily on data, and data are often represented in graphics. The media use graphics often to illustrate articles and make points more convincingly. Reading and understanding this kind of information therefore is an essential component of *mathematical literacy*.

Content area: *Uncertainty* – the focus is on exploratory data analysis. The mathematical content is restricted to reading data from a bar graph or pie chart.

Competency cluster: *Reproduction* – interpret and recognise situations in contexts that require no more than direct inference. Students need to solve the problem by decoding and interpreting a familiar, practised standard representation of a well known mathematical object.

The question requires students to:

- Follow the written instructions.
- Decide which of the two graphs is relevant.
- Locate the correct information in that graph.

Successful students answered either "27.1 million zeds" or "27 100 000 zeds" or even just "27.1" without the unit "zeds". Students were most successful on this question in France (92%), the Netherlands (91%), Canada (90%), the partner country Liechtenstein (89%), Belgium, Portugal and Finland (88%). On average across OECD countries only 7% of students failed to respond to this question.

Note that there is another question in this unit (EXPORTS – Question 2) and this is presented in the section *Examples of moderate to difficult questions in PISA 2003 mathematics*.

THE BEST CAR

A car magazine uses a rating system to evaluate new cars, and gives the award of "The Car of the Year" to the car with the highest total score. Five new cars are being evaluated, and their ratings are shown in the table.

Car	Safety Features (S)	Fuel Efficiency (F)	External Appearance (E)	Internal Fittings (T)
Ca	3	1	2	3
M2	2	2	2	2
Sp	3	1	3	2
N1	1	3	3	3
KK	3	2	3	2

The ratings are interpreted as follows:

3 points = Excellent

2 points = Good

1 point = Fair

Question 1: THE BEST CAR

To calculate the total score for a car, the car magazine uses the following rule, which is a weighted sum of the individual score points:

$$\text{Total Score} = (3 \times S) + F + E + T$$

Calculate the total score for Car "Ca". Write your answer in the space below.

Total score for "Ca": ....

THE BEST CAR – Question 1 illustrates Level 2 in PISA 2003 mathematics, with a difficulty of 447 PISA score points. On average across OECD countries, 73% of students were able to do this successfully.

Context: *Public* – an article in a car magazine is a very familiar context, especially for males. The underlying mathematics is relevant for males and females as everyone is presented with this kind of problem, that is, the evaluation of a consumer good using a rating system, whether it be cars, washing machines, coffee makers, etc. This is therefore an important part of *mathematical literacy*.

Content area: *Change and relationships* – the focus is on the relationship of numbers in a formula

Competency cluster: *Reproduction* – students need to reproduce a practised procedure. However, this is not trivial as it involves an equation and students typically find it difficult to work with equations presented in such a real-world context.

The question requires students to:

- Read and understand a relatively straightforward question.
- Multiply a number by 3.
- Add four simple numbers.

Successful students answered "15 points". Students were most successful on this question in the partner economy Macao-China (90%), the partner countries/economies Liechtenstein and Hong Kong-China (87%), Korea (84%), Canada (82%) and Denmark, Austria and Japan (80%). On average across OECD countries, 10% of students did not respond to this question, but this included 24% of students in Mexico, 23% in the partner country Brazil and 21% of students in Greece. It is interested to note that the OECD average for female students was 74.5%, while that for male students was 71.33% – a significant difference favouring female students!

Note that there is another question in this unit (THE BEST CAR – Question 2) and this is presented in the section *Examples of difficult questions in PISA 2003 mathematics*.

SKATEBOARD

Eric is a great skateboard fan. He visits a shop named SKATERS to check some prices.

At this shop you can buy a complete board. Or you can buy a deck, a set of 4 wheels, a set of 2 trucks and a set of hardware, and assemble your own board.

The prices for the shop's products are:

Product	Price in zeds	
Complete skateboard	82 or 84	
Deck	40, 60 or 65	
One set of 4 Wheels	14 or 36	
One set of 2 Trucks	16	
One set of hardware (bearings, rubber pads, bolts and nuts)	10 or 20	

Question 1: SKATEBOARD

Eric wants to assemble his own skateboard. What is the minimum price and the maximum price in this shop for self-assembled skateboards?

(a) Minimum price: .... zeds.

(b) Maximum price: .... zeds.

SKATEBOARD – Question 1 illustrates two different levels of proficiency depending on whether students gave a fully or partially correct answer. Here, a partially correct answer scored at 1 point illustrates a Level 2 performance with a difficulty of 464 PISA score points. A fully correct answer illustrates Level 3 with a difficulty of 496 PISA score points. On average across OECD countries, 11% of students were only capable of achieving the partial 1 point level.

Context: Personal – skateboards are part of the youth culture; either students skateboard themselves or watch others do it – especially on television.

Content area: Quantity – the students are asked to find a minimum and maximum price for the construction of a skateboard, under given numerical conditions. The skills needed to solve this problem are certainly an important part of *mathematical literacy* as they make it possible to make more informed decisions in daily life.

Competency cluster: Reproduction – students need to solve the problem by finding a simple strategy to produce the minimum and maximum and reproduce practised knowledge in combination with the performance of a routine addition.

The question requires students to:

- Interpret the question correctly and so understand that they need to provide two answers.
- Extract the relevant information from a simple table.
- Find a simple strategy to come up with the minimum and maximum (this is simple because the strategy that seems trivial actually works: for the minimum take the lower numbers, for the maximum the larger ones).
- Perform a basic addition. (The whole number addition: 40 + 14 + 16 + 10 equals 80, gives the minimum, and the maximum is found by adding the larger numbers: 65 + 36 + 16 + 20 = 137).

This question illustrates Level 2 when the students give a partially correct answer: by giving either the minimum or the maximum, but not both. On average across the OECD countries 11% of students gave a partially correct answer. This was the case for 28% of students in France and 13% of students in Mexico, Luxembourg and the partner country Serbia.

On average across the OECD countries, the majority of students responded answer this question – with only 5% failing to do so. (Although this was 12% of students in Turkey and 11% of students in Greece and Japan).

Note that full credit for SKATEBOARD Q1 displays mathematical proficiency at Level 3 is discussed for this performance in the section *Examples of moderate to difficult questions in PISA 2003 mathematics*.

CUBES

Question 1: CUBES

In this photograph you see six dice, labelled (a) to (f). For all dice there is a rule:

The total number of dots on two opposite faces of each die is always seven.

*Write in each box the number of dots on the **bottom** face of the dice corresponding to the photograph.*

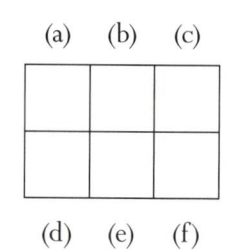

(a)	(b)	(c)

(d)　(e)　(f)

CUBES – Question 1 illustrates Level 2 in PISA 2003 mathematics with a difficulty of 478 PISA score points. On average across OECD countries, 68% of students were able to do this successfully.

Context: *Educational and occupational* – the context in this case is somewhat difficult to classify: for many students the number cubes are very familiar recreational objects and therefore the context could be classified as "personal", but these objects are also frequently seen in school contexts. Further, the question calls for spatial representation skills that are required at a basic level in many occupations.

Content area: *Space and shape* – spatial representation

Competency cluster: *Reproduction* – apply a simple given rule and use basic spatial representation skills. Even if students are not familiar with number cubes (or dice) the essential rule is stated clearly in the introductory text. These competencies are essential to mathematical literacy, but are only required a very basic level here.

The question requires students to:

- Apply the rule, that the opposites sum up to 7, six times.
- Use spatial representation skills to 'transfer' the presented photo into the table.

Successful students answered "Top row (1 5 4) Bottom Row (2 6 5)" or drew a diagram showing the correct numbers on the faces of the cubes. Students were most successful on this question in Finland and Switzerland (80% correct), Japan (79%), Sweden (78%) and France, Canada and the partner country Liechtenstein (76%). Most students across the OECD countries responded to this question – only 6% failed to do so, although this was 12% of students in Hungary, Greece and in the partner countries Serbia, Tunisia and Brazil.

So, what characteristics do these examples of easier mathematics questions share beyond the predominance of questions in the *reproduction* competency cluster? First, the response formats used in this set of easier questions are similar. All eight questions require students to undertake rather convergent thinking and provide a simple, short and rather closed constructed response, usually a single numeric answer. None of these released questions requires students to write an explanation of their solution, or a justification of their result. All of the released easy questions involve rather directed instructions towards finding a single correct numeric answer, with little reasoning required. In general, it is observed that questions of these forms are the easiest to answer. They place no demands on students in relation to deciding what kind of response would constitute an answer to the question asked. Students can find or calculate the answer, or they cannot. In most cases, the question formats even indicated where or how the student should respond.

Second, there is a low level of complexity in the language used in questions and the unit or stimulus they are related to. Most of the easy questions have relatively low reading demands. The text in the stimulus of each unit generally consists of simple, direct statements. Similarly, the questions are relatively short and direct. The language demand of questions can be an important factor in determining the difficulty of questions and is discussed in more detail in Chapter 5. Finally, the graphical or pictorial displays in the setting of the units are also of familiar formats and ones that students would have had experience in creating or manipulating in school or life situations.

EXAMPLES OF MODERATE TO DIFFICULT MATHEMATICS QUESTIONS FROM PISA 2003

Forty-three of the mathematics questions used in the PISA 2003 assessment lie on that part of the *mathematical literacy* scale covered by Levels 3 and 4, and can therefore be regarded as representing intermediate levels of difficulty. Eighteen of these questions, coming from 15 different units, have been released. These are listed together with information about various characteristics of the questions in Table 3.2. Figures showing country level performances on many of these moderate to difficult items are found in Annex A1, Figures A.1.9 through A.1.23.

Around this middle part of the reported PISA *mathematical literacy* scale, the increased difficulty of questions relative to those in Levels 1 and 2 can be seen in the increased number of questions from the *connections* competency cluster, and the appearance of questions from the *reflection* competency cluster. Mathematics questions in these competency clusters typically have greater complexity and impose increased cognitive demands. All four of the mathematical content areas are represented among the released questions from these levels. Similarly, all contexts are represented. And unlike the easier questions discussed earlier, there is a mix of different response formats. Almost half of these released questions are of the short answer type, similar to the entire set of easier questions, but there are also multiple-choice questions and questions requiring an extended response. It is also evident that these questions impose a greater reading load than those in Levels 1 and 2.

Table 3.2
Characteristics of moderate to difficult released questions

Item Code	Question	OECD average percent correct	Location on PISA scale (PISA score points) Question	Full/partial credit points 1	2	3	Traditional topic	Content area ("Overarching Idea")	Competency cluster	Context ("Situation")	Length of question[1]	Response format
M806Q01	STEP PATTERN – Question 1	66	484	484			Number	Quantity	Reproduction	Educational and occupational	Short	Short Answer
M520Q01	SKATEBOARD – Question 1	72	480	464	496		Number	Quantity	Reproduction	Personal	Long	Short Answer
M484Q01	BOOKSHELVES – Question 1	61	499	499			Number	Quantity	Connections	Educational and occupational	Medium	Short Answer
M555Q02	NUMBER CUBES – Question 2	63	503	503			Geometry	Space and shape	Connections	Personal	Long	Complex Multiple Choice
M150Q02[2]	GROWING UP – Question 2	69	472	420	525		Data	Change and relationships	Reproduction	Scientific	Medium	Short Answer
M402Q01	INTERNET RELAY CHAT – Question 1	54	533	533			Measurement	Change and relationships	Connections	Personal	Medium	Short Answer
M467Q01	COLOURED CANDIES – Question 1	50	549	549			Data	Uncertainty	Reproduction	Personal	Short	Multiple Choice
M505Q01	LITTER – Question 1	52	551	551			Data	Uncertainty	Reflection	Scientific	Medium	Extended Response
M520Q03	SKATEBOARD – Question 3	50	554	554			Number	Quantity	Connections	Personal	Long	Short Answer
M468Q01	SCIENCE TESTS – Question 1	47	556	556			Data	Uncertainty	Reproduction	Educational and occupational	Medium	Short Answer
M509Q01	EARTHQUAKE – Question 1	46	557	557			Data	Uncertainty	Reflection	Scientific	Long	Multiple Choice
M510Q01	CHOICES – Question 1	49	559	559			Number	Quantity	Connections	Educational and occupational	Medium	Short Answer

A Question of Difficulty: Questions from PISA 2003

Table 3.2
Characteristics of moderate to difficult released questions *(continued)*

Item Code	Question	OECD average percent correct	Location on PISA scale (PISA score points) Question	Full/partial credit points 1	2	3	Traditional topic	Content area ("Overarching Idea")	Competency cluster	Context ("Situation")	Length of question[1]	Response format
M438Q02	EXPORTS – *Question 2*	48	565	**565**			Number	*Uncertainty*	*Connections*	*Public*	Medium	Multiple Choice
M520Q02	SKATEBOARD – *Question 2*	46	570	**570**			Number	*Quantity*	*Reproduction*	*Personal*	Short	Multiple Choice
M150Q03	GROWING UP – *Question 3*	45	574	**574**			Data	*Change and relationships*	*Connections*	*Scientific*	Medium	Extended Response
M179Q01	ROBBERIES – *Question 1* Partial 1 Point	30	635	577	694		Data	*Uncertainty*	*Connections*	*Public*	Short	Extended Response
M413Q03	EXCHANGE RATE – *Question 3*	40	586	**586**			Number	*Quantity*	*Reflection*	*Public*	Medium	Extended Response
M124Q03	WALKING – *Question 3* Partial 1 point	21	665	**605**	666	723	Algebra	*Change and relationships*	*Connections*	*Personal*	Medium	Extended Response

1. Short questions contain fewer than 50 words. Medium-length questions contain 51 to 100 words. Long questions contain more than 100 words. Length of question in relation to question difficulty is discussed in detail in Chapter 5.

STEP PATTERN

Question 1: STEP PATTERN

Robert builds a step pattern using squares. Here are the stages he follows.

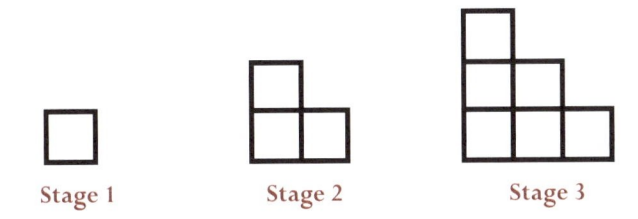

Stage 1   Stage 2   Stage 3

As you can see, he uses one square for Stage 1, three squares for Stage 2 and six for Stage 3.

How many squares should he use for the fourth stage?

Answer: .... squares.

STEP PATTERN – Question 1 illustrates Level 3 in PISA 2003 mathematics, with a difficulty of 484 PISA score points. On average across OECD countries, 66% of students were able to do this successfully.

Context: *Educational and occupational* – This problem would be representative of similar tasks seen commonly in mathematics classes or textbooks; indeed it is almost a pure mathematical problem. The importance of such a question in a test focusing on *mathematical literacy* is not immediately clear. Such questions are not seen in newspapers, on television, or at work. But recognising regularities or patterns and being able to predict the next member of the sequence are helpful skills when structured processing is required. It is a well known fact that problems like these appear in many psychological tests. Some mathematicians do not approve of questions that ask for the next member of a given string of integers, as it can be argued mathematically that any answer is correct. For students of this age this turns out, in practice, not to be a problem and certainly is not for this particular question since it presents a pattern with both a numeric and geometric base.

Content area: *Quantity* – recognising a pattern from a numeric and geometric base.

Competency cluster: *Reproduction* – use of very basic strategies and no need for mathematisation. The question is simple and clearly stated and it is not strictly necessary to read the text. There are at least two simple possible strategies: Either count the numbers of each object (1, 3, 6 and the next number will be 10); or sketch the next object and then count the number of squares.

Successful students answered "10". Students were most successful on this question in Japan (88%), the partner economy Hong Kong-China (83%), Korea and the partner economy Macao-China (80%) and the Czech Republic, Denmark and Norway (78%). This was a question that the majority of students responded to – on average across OECD countries only 1% failed to do so and this did not surpass 5% in any of the OECD countries.

SKATEBOARD

Eric is a great skateboard fan. He visits a shop named SKATERS to check some prices.

At this shop you can buy a complete board. Or you can buy a deck, a set of 4 wheels, a set of 2 trucks and a set of hardware, and assemble your own board.

The prices for the shop's products are:

Product	Price in zeds	
Complete skateboard	82 or 84	
Deck	40, 60 or 65	
One set of 4 Wheels	14 or 36	
One set of 2 Trucks	16	
One set of hardware (bearings, rubber pads, bolts and nuts)	10 or 20	

Question 1: SKATEBOARD

Eric wants to assemble his own skateboard. What is the minimum price and the maximum price in this shop for self-assembled skateboards?

(a) Minimum price: .... zeds.

(b) Maximum price: .... zeds.

SKATEBOARD – Question 1 illustrates two different levels of proficiency depending on whether students gave a fully or partially correct answer. A partially correct answer illustrates Level 2 and has been discussed earlier. A fully correct answer illustrates Level 3 with a difficulty of 496 PISA score points. On average across OECD countries, 72% of students were able to do this successfully.

Context: *Personal* – skateboards are part of the youth culture; either students skateboard themselves or watch others do it – especially on television.

Content area: *Quantity* – the students are asked to find a minimum and maximum price for the construction of a skateboard, under given numerical conditions. The skills needed to solve this problem are certainly an important part of *mathematical literacy* as they make it possible to make more informed decisions in daily life.

Competency cluster: *Reproduction* – students need to solve the problem by finding a simple strategy to produce the minimum and maximum and reproduce practised knowledge in combination with the performance of a routine addition.

The question requires students to:

- Interpret the question correctly and so understand that they need to provide two answers.
- Extract the relevant information from a simple table.
- Find a simple strategy to come up with the minimum and maximum (this is simple because the strategy that seems trivial actually works: for the minimum take the lower numbers, for the maximum the larger ones).
- Perform a basic addition. (The whole number addition: 40 + 14 + 16 + 10 equals 80, gives the minimum, and the maximum is found by adding the larger numbers: 65 + 36 + 16 + 20 = 137).

Students were most successful on this question, providing both the minimum (80) and the maximum (137), in Finland (81%), the partner country Liechtenstein and Switzerland (76%), Canada (75%), Australia, New Zealand, Belgium and Austria (74%).

On average across the OECD countries, the majority of students responded to this question – with only 5% failing to do so (although this was 12% of students in Turkey and 11% of students in Greece and Japan).

Question 2: SKATEBOARD

The shop offers three different decks, two different sets of wheels and two different sets of hardware. There is only one choice for a set of trucks.

How many different skateboards can Eric construct?

A 6
B 8
C 10
D 12

SKATEBOARD – Question 2 illustrates Level 4 in PISA 2003 mathematics, with a difficulty of 570 score points. On average across OECD countries, 46% of students were able to do this successfully.

Context: *Personal*

Content area: *Quantity* – routine computation. The skills needed to solve this problem are certainly an important part of *mathematical literacy* as they make it possible to make more informed decisions in daily life.

Competency cluster: *Reproduction* – all the required information is explicitly presented. Students need to understand what the required "strategy" is and then carry out that strategy. For students having identified the required strategy, the mathematics involves the basic routine computation: 3 x 2 x 2 x 1. However, if students do not have experience with such combinatorial calculations, their strategy might involve a systematic listing of the possible combinatorial outcomes. There are well-known algorithms for this (such as a tree diagram). The strategy to find the number of combinations can be considered as common, and more or less routine. It involves following and justifying standard quantitative processes, including computational processes, statements and results.

The question requires students to:

- Interpret correctly a text in combination with a table.
- Apply accurately a simple enumeration algorithm.

Successful students answered "D" (12). Students were most successful on this question in Japan (67%), Korea (65%), Denmark and the partner economy Hong Kong-China (60%).

The incorrect answer most frequently given by students across the OECD countries was "A" (25%), followed by "B" (18%). Only in Korea and Hungary did more students chose answer "B" than "A", and in Japan and the Netherlands students were equally divided among these two incorrect categories. To get answer "B" students may have added the whole numbers in the question to get a total of 8. To get answer "A" it is most likely that students misread the question and missed one of the components with two different sets (either the wheels or the hardware).

Question 3: SKATEBOARD

Eric has 120 zeds to spend and wants to buy the most expensive skateboard he can afford.

How much money can Eric afford to spend on each of the 4 parts? Put your answer in the table below.

Part	Amount (zeds)
Deck	
Wheels	
Trucks	
Hardware	

SKATEBOARD – Question 3 illustrates the lower part of Level 4, with a difficulty of 554 PISA score points (ten points above the boundary with Level 3). On average across OECD countries, 50% of students were able to do this successfully.

Context: Personal

Content area: Quantity – students are asked to compute what is the most expensive Skateboard that can be bought for 120 zeds by using some kind of quantitative process that has not been described. But this task is certainly not straightforward: there is no standard procedure or routine algorithm available.

Competency cluster: Connections – students need to use an independent and not routine problem-solving approach. Students may use different strategies in order to find the solution, including trial and error. So the setting of this problem is familiar or quasi-familiar but the problem to be solved is not simply routine. Students have to pose a question (How do we find…?), look at the table with prices, make combinations and do some computation. A strategy that will work with this problem is to first use all the higher values, and then adjust the answer, reducing the price until the desired maximum of 120 zed is reached. So taking the deck at 65 zed, the wheels at 36 zed, the trucks at 16 zed (no choice here) and the hardware at 20 zed. This gives a total of 137 zed – the maximum found earlier in Question 1. The cost needs to be reduced by at least 17 zed. It is possible to reduce the cost by 5 zed or 25 zed on the deck, by 22 zed on the wheels or by 10 zed on the hardware. The best solution is clear: save 22 zed on the wheels.

The question requires students to:

- Reason in a familiar context.
- Connect the question with the data given in the table, or in other words, relate text-based information to a table representation.
- Apply a non-standard strategy.
- Carry out routine calculations.

Successful students answered "65 zeds on a deck, 14 on wheels, 16 on trucks and 20 on hardware". Students were most successful on this question in the partner economy Macao-China (65%), the partner economy Hong Kong-China (62%), Finland, Canada and Sweden (59%), Belgium (58%) and Australia (57%). Seventeen percent of students across the OECD countries narrowly missed the correct answer and only gave correct prices for three of the four parts and this was as much as 27% of students in the partner country Thailand, 26% in Mexico, 21% in the partner country Serbia, and 20% in Luxembourg, the United States and Greece.

BOOKSHELVES

Question 1: BOOKSHELVES

To complete one set of bookshelves a carpenter needs the following components:

- 4 long wooden panels,
- 6 short wooden panels,
- 12 small clips,
- 2 large clips and
- 14 screws.

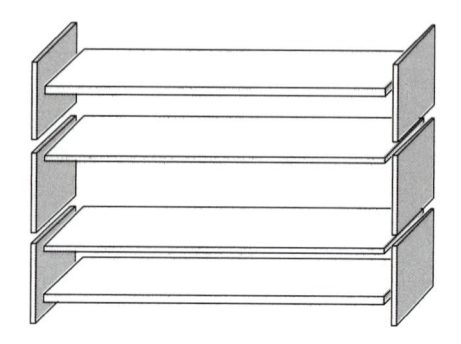

The carpenter has in stock 26 long wooden panels, 33 short wooden panels, 200 small clips, 20 large clips and 510 screws.

How many sets of bookshelves can the carpenter make?

Answer: ....

BOOKSHELVES – Question 1 illustrates Level 3 in PISA 2003 mathematics, with a difficulty of 499 PISA score points. On average across the OECD countries, 61% of students were able to do this successfully.

Context: *Educational and occupational* – The stem uses both a visualisation as well as text, with a lot of numbers, and a clear and short question. Almost by definition problems from an occupational context fit well with *mathematical literacy*. The problem in principle also has a certain level of authenticity: it stands for large collection of problems that have as core the attribution of parts to a production process in order to optimise the quantities of required components and to minimise waste.

Content area: *Quantity* – computation of ratios. Students compute the following ratio for each of the components: available components/required components per set of bookshelves. This gives: 26/4 (for long panels); 33/6 (for short panels); 200/12 (for small clips); 20/2 (for large clips); 510/14 (for screws).

Competency cluster: *Connections* – strategic thinking and some mathematisation. Students analyse the ratios they have computed to find that the smallest answer is 33/6, or 5.5. However, this is only an indication of the solution to the problem and 5.5 would not be a satisfactory response to the question asked. Students need to interpret this mathematical answer back into the bookshelves context to find the correct real-world solution: 5 sets of bookshelves.

The question requires students to:

- Develop a strategy to connect two bits of information for each component: the number available, and the number needed per set of bookshelves.
- Use logical reasoning to link that analysis across the components to produce the required solution.
- Communicate the mathematical answer as a real-world solution.

Successful students answered "5". Students were most successful on this question in Finland and the partner economy Hong Kong-China (74%), Korea, the Czech Republic, Belgium and Denmark (72%). On average across OECD countries 29% of students responded to this question but gave an incorrect answer and 10% did not respond at all.

NUMBER CUBES

Question 2: NUMBER CUBES

On the right, there is a picture of two dice.

Dice are special number cubes for which the following rule applies:

The total number of dots on two opposite faces is always seven.

You can make a simple number cube by cutting, folding and gluing cardboard. This can be done in many ways. In the figure below you can see four cuttings that can be used to make cubes, with dots on the sides.

Which of the following shapes can be folded together to form a cube that obeys the rule that the sum of opposite faces is 7? For each shape, circle either "Yes" or "No" in the table below.

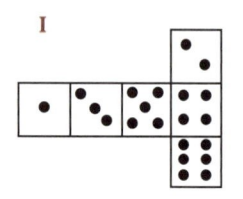

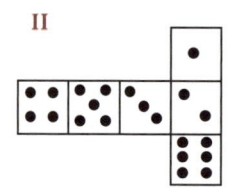

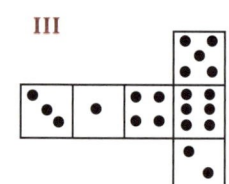

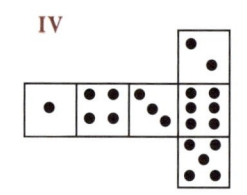

Shape	Obeys the rule that the sum of opposite faces is 7?
I	Yes / No
II	Yes / No
III	Yes / No
IV	Yes / No

NUMBER CUBES – Question 2 illustrates Level 3 in PISA 2003 mathematics, with a difficulty of 503 PISA score points. On average across OECD countries, 63% of students were able to do this successfully.

Context: *Personal* – many games that children encounter during their education, whether formal or informal, use number cubes. The problem does not assume any previous knowledge about this cube, in particular the rule of construction: two opposite sides have a total of seven dots.

Content area: *Space and shape* – spatial reasoning skills. The given construction rule emphasises a numerical aspect, but the problem posed requires some kind of spatial insight or mental visualisation technique. These competencies are an essential part of *mathematical literacy* as students live in three-dimensional space, and often are confronted with two-dimensional representations. Students need to mentally imagine the four plans of number cubes reconstructed into a three-dimensional number cube and judge whether they really obey the numerical construction rule.

Competency cluster: *Connections* – The problem is certainly not routine: students need to connect written information, graphical representation and interpret back-and-forth. However, all the relevant information is clearly presented in writing and with graphics.

The question requires students to:

- Encode and interpret spatially two-dimensional objects.
- Interpret the connected three-dimensional objects.
- Interpret back-and-forth between model and reality.
- Check certain basic quantitative relations.

Successful students answered "No, Yes, Yes, No" in that order. Students were most successful on this question in Japan (83% correct), Korea (81%), Finland (76%), Belgium (74%), the Czech Republic and Switzerland (73%). A further 16% of students on average across OECD countries narrowly missed the fully correct answer and provided three out of four of the correct shapes and only 2% of students did not respond to the question.

GROWING UP

Youth grows taller

In 1998 the average height of both young males and young females in the Netherlands is represented in this graph:

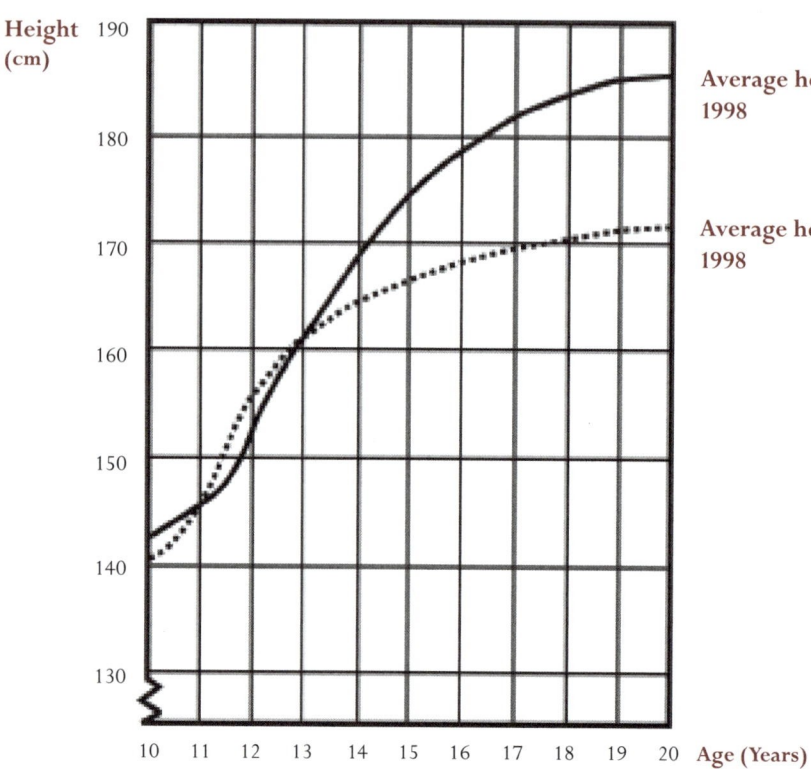

Average height of young males 1998

Average height of young females 1998

Question 2: GROWING UP

According to this graph, on average, during which period in their life are females taller than males of the same age?

....

GROWING UP – Question 2 illustrates two different levels of proficiency depending on whether students gave a fully or partially correct answer. A partially correct answer illustrates exactly the boundary between Level 1 and Level 2 with a difficulty of 420 PISA score points. Here a fully correct answer illustrates Level 3 with a score of 2 points for PISA scale difficult of 525 score points. On average across OECD countries, 69% of students were able to do this successfully.

Context: Scientific

Content area: Change and relationships – focus on the relationship between age and height. The mathematical content can be described as belonging to the "data" domain: the students are asked to compare characteristics of two data sets, interpret these data sets and draw conclusions.

Competency cluster: Reproduction – interpret and decode reasonably familiar and standard representations of well known mathematical objects. Students need to think and reason (where do the graphs have common points?), use argumentation to explain which role these points play in finding the desired answer and communicate and explain the argumentation. However, all these competencies essentially involve reproduction of practised knowledge.

The question requires students to:

- Interpret and use a graph.
- Make conclusions directly from a graph.
- Report the results of their reasoning in a precise manner.

Students who were most successful on this question showed that their reasoning and/or insight was well directed and properly identified the continuum from 11 to 13 years. This was the case for 80% of students in Korea, 74% in the partner country Liechtenstein, 72% in France, 69% in Belgium and 67% in the Netherlands and Finland. Across countries, the majority of successful students communicated the correct interval as follows: "Between age 11 and 13"; "From 11 years old to 13 years old, girls are taller than boys on average"; or "11-13". However, a minority of successful students stated the actual years when girls are taller than boys, which is correct in daily-life language: "Girls are taller than boys when they are 11 and 12 years old"; or "11 and 12 years old." This concerned only 5% or less of the fully correct answers given in 16 of the OECD countries and this only surpassed 10% of fully correct answers in Turkey (21%), Mexico (19%) and Ireland (13%).

Seven percent of students on average across the OECD countries did not respond to this question. This concerned less than 3% of students in the Netherlands, Finland, Canada and the partner economy Macao-China.

INTERNET RELAY CHAT

Mark (from Sydney, Australia) and Hans (from Berlin, Germany) often communicate with each other using "chat" on the Internet. They have to log on to the Internet at the same time to be able to chat.

To find a suitable time to chat, Mark looked up a chart of world times and found the following:

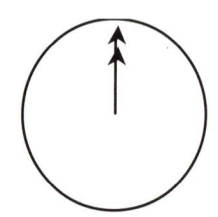

Greenwich 12 Midnight   Berlin 1:00 AM   Sydney 10:00 AM

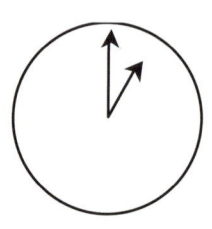

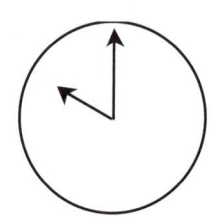

Question 1: INTERNET RELAY CHAT

At 7:00 PM in Sydney, what time is it in Berlin?

Answer: ....

INTERNET RELAY CHAT – Question 1 illustrates Level 4 with a difficulty of 533 PISA score points. On average across OECD countries 54% of students were able to do this successfully.

Context: Personal – this assumes either that students are familiar at some level with chatting over the internet, and/or they know about time differences in this or another context.

Content area: Change and relationships

Competency cluster: Connections – solving a non-routine problem, using simple mathematical tools, and making use of different representations. The problem does need some mathematisation, starting with identifying the relevant mathematics. The question is simple, and so are the numbers and the actual operations needed (adding and subtracting whole numbers). So the complexity lies really in the mathematisation: first the students have to identify the time difference between Berlin and Sydney (9 hours). Then they have to appreciate the fact that it is 9 hours later in Sydney. Then they have to apply this difference to the new situation.

This question requires students to:

- Identify the relevant mathematics.
- Solve a non-routine, but simple problem.
- Use different representations.

Successful students answered "10 AM or 10:00". At least 60% of students answered this question correctly in the Czech Republic, Denmark, France, Belgium, Germany, Switzerland, Luxembourg, Korea, Japan, the Slovak Republic and Austria, as well as in the partner country Liechtenstein.

Nearly all students tried to respond to this question; across the OECD on average only 4% of students failed to respond.

Note that this unit includes one other question (INTERNET RELAY CHAT – Question 2) and this is presented in the section *Examples of difficult questions in PISA 2003 mathematics*.

COLOURED CANDIES

Question 1: COLOURED CANDIES

Robert's mother lets him pick one candy from a bag. He can't see the candies. The number of candies of each colour in the bag is shown in the following graph.

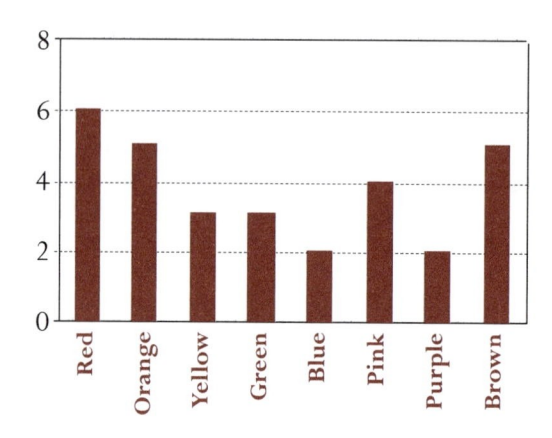

What is the probability that Robert will pick a red candy?

A 10%

B 20%

C 25%

D 50%

COLOURED CANDIES – Question 1 illustrates Level 4 in PISA 2003 mathematics and has a difficulty of 549 PISA score points. On average across the OECD countries, 50% of students were able to do this successfully.

Context: *Personal* – Many students can relate to this context through previous experiences, and such experiences involve some kind of probabilistic reasoning as young children do prefer certain colours or flavours. And they realise that certain colours are less abundant than others. Perhaps the problem lacks some authenticity for students at age 15, but the underlying concepts are valuable and relevant.

Content area: *Uncertainty* – this problem represents a wide array of problems that involve some thinking about chance. This problem measures an important aspect of *mathematical literacy* through its presentation of a more or less realistic situation that elicits probabilistic thinking, and its demand that students make direct and explicit connections between the context and a standard mathematical representation of a key aspect of the context – namely a bar chart representing the frequency distribution by colour of candies in the bag. This question formalises dealing with uncertainty in a fairly straightforward way.

Competency cluster: *Reproduction* – a complex and demanding combination of individual Reproduction competencies

The question requires students to:

- Identify relevant information from the graph (there are 6 red candies).
- Identify and calculate from the graph the total number of candies (6 + 5 + 3 + 3 + 2 + 4 + 2 + 5 or altogether 30 candies).
- Produce a basic probability calculation to get to the answer: 6 out of 30 is 20%.

Successful students answered "B" (20%). Students were most successful on this question in Iceland (76%), Korea (73%), the partner economy Hong Kong-China (72%), the Netherlands (69%) and Denmark (66%).

On average across the OECD countries, only 2% of students did not respond to this question.

LITTER

Question 1: LITTER

For a homework assignment on the environment, students collected information on the decomposition time of several types of litter that people throw away:

Type of Litter	Decomposition time
Banana peel	1–3 years
Orange peel	1–3 years
Cardboard boxes	0.5 year
Chewing gum	20–25 years
Newspapers	A few days
Polystyrene cups	Over 100 years

A student thinks of displaying the results in a bar graph.

Give **one** *reason why a bar graph is unsuitable for displaying these data.*

LITTER – Question 1 illustrates Level 4 in PISA 2003 mathematics, with a difficulty of 551 PISA score points. On average across OECD countries 52% of students were able to do this successfully.

Context: Scientific

Content area: Uncertainty – this question aims to test whether students are able to reason correctly about how to represent numbers (data) appropriately and their skills to effectively communicate this. As such, the scoring of this question is important. There are two possible correct answers, but students only need to give one of these: Either students make an argument based on the large differences in magnitude of the numbers involved, and the resulting difficulty in displaying these; or students make an argument based on the variability of the data within the different categories, and the resulting uncertainty in constructing a display.

Competency cluster: Reflection – Visualising an argument or data in an appropriate, meaningful and convincing way, and conversely, judging such representations on their qualities are key aspects of mathematical literacy. This requires some kind of reflection on the available data.

The question requires students to:

- Interpret the data.
- Reflect on the data.
- Communicate the results of their reflection.

Successful students gave answers focusing on either the big variance in data or the variability of the data for some categories. The question places an emphasis on communication of results. An examination of student responses to this question illustrates this point. The following two student responses were scored as correct answers:

- "You will get a mess, one starts at 0.5 years and another one at more than hundred years."
- "You have to make a vertical axis that goes minimally to 100 years with small steps because you need to able to read "a couple of days"."

Other correct answers could include:

- "The length of the bar for "polystyrene cups" is undetermined."
- "You cannot make one bar for 1–3 years or one bar for 20–25 years."
- "The difference in the lengths of the bars of the bar graph would be too big."
- "If you make a bar with length 10 centimetres for polystyrene, the one for cardboard boxes would be 0.05 centimetres."

Students were most successful on this question in Korea (74% correct), Finland (73%), Iceland (71%), the partner economy Hong Kong-China (68%) and Norway (67%). On average across OECD countries 32% of students attempted to answer the question but gave an incorrect answer. This varied from less than 20% of students in Korea and Poland to 76% of students in the United States. Examples of incorrect answers include:

- "Because it will not work."
- "A pictogram is better."
- "You cannot verify the info."
- "Because the numbers in the table are only approximations."

Across the OECD countries on average, 16% of students did not respond to this question.

SCIENCE TESTS

Question 1: SCIENCE TESTS

In Mei Lin's school, her science teacher gives tests that are marked out of 100. Mei Lin has an average of 60 marks on her first four Science tests. On the fifth test she got 80 marks.

What is the average of Mei Lin's marks in Science after all five tests?

Average: ………………………

SCIENCE TESTS – Question 1 illustrates Level 4 in PISA 2003 mathematics, with a difficulty of 556 PISA score points. On average across the OECD countries, 47% of students were able to do this correctly.

Context: *Educational and occupational* – this is a very familiar context for many students.

Content area: *Uncertainty* – weighted average. The problem-solving process could be as follows: add the score of 80 marks to the existing average for the first four science tests, that is, 60 marks. So: 60 + 60 + 60 + 60 + 80 = 320. (Or: 4 x 60 plus 80). Then divide this number by 5 to get the answer of 64 marks.

By far the most common incorrect response to this question was the answer 70. It is clear that this answer is incorrect, and it seems plausible to assume that these students have not read the stem of the problem accurately enough and rushed to the conclusion that the requested answer was the simple average of 60 and 80 (calculated as 60 + 80 divided by 2) rather than a weighted average that recognises that the total of the first four test scores must be 240.

Competency cluster: *Reproduction* – the concept of average is tested by giving a problem with a very familiar context, with simple numbers. The fact that the average of the first four scores were given might have added to the complexity, as the frequency of the incorrect answer 70 suggests.

The question requires students to:

- Read carefully.
- Have a proper understanding of the mathematical concept of the "average".
- "Reverse engineer" the rule for calculating an average to find the new average. This involves both the mathematisation of the concept of average and mathematical manipulation of the result.

Successful students answered "64". The most successful students on this question were in the partner economy Hong Kong-China (75% correct), the partner economy Macao-China (69%), Korea (67%), Japan (63%) and Canada (60%).

Across the OECD countries on average 16% of students did not respond to this question.

EARTHQUAKE

Question 1: EARTHQUAKE

A documentary was broadcast about earthquakes and how often earthquakes occur. It included a discussion about the predictability of earthquakes.

A geologist stated: "In the next twenty years, the chance that an earthquake will occur in Zed City is two out of three."

Which of the following best reflects the meaning of the geologist's statement?

A ⅔ × 20 = 13.3, so between 13 and 14 years from now there will be an earthquake in Zed City.

B ⅔ is more than ½, so you can be sure there will be an earthquake in Zed City at some time during the next 20 years.

C The likelihood that there will be an earthquake in Zed City at some time during the next 20 years is higher than the likelihood of no earthquake.

D You cannot tell what will happen, because nobody can be sure when an earthquake will occur.

EARTHQUAKE – Question 1 illustrates Level 4 in PISA 2003 mathematics, with a difficulty of 557 PISA score points. On average across OECD countries, 46% if students were able to do this successfully.

Context: *Scientific*

Content area: *Uncertainty* – statistical forecasting/predictions. This question illustrates an important part of *mathematical literacy*. Experts often make predictions, although these are seldom transparent or explicit. For example, expressions used to forecast the weather, such as "there is a 20% chance of rain tomorrow". The viewer or reader thinks that there is a good chance it will remain dry tomorrow, but cannot complain if it rains for a substantial part of the day. Intelligent and mathematically literate citizens should be able to reflect in a critical way on what is actually meant by such a prediction.

Competency cluster: *Reflection* – students need to consider a given statement and reflect upon the meaning of that statement and four possible responses. Ideally such a question would require students to explain the result of their reflection in their own words, but such answers would probably be difficult to score objectively. Therefore the format of multiple-choice has been chosen. This means an extra step for the students: they may reflect first, and try to connect the result of this process to one of the four possible responses. Alternatively, students may consider the four possible responses and try to judge which one is the most likely. In this case the Reflection process takes a slightly different form.

Successful students answered "C". Students were most successful on this question in Japan (68%), Korea (64%) and Finland and New Zealand (59%). It is interesting that quite a large number of students chose the wrong answer "D" – 22% on average across OECD countries and as many as 39% of students in the Slovak Republic and the partner country Serbia and 36% in the Czech Republic. This statement is arguably correct but is not an answer to the question asked.

On average across OECD countries 9% of students did not respond to this question.

CHOICES

Question 1: CHOICES

In a pizza restaurant, you can get a basic pizza with two toppings: cheese and tomato. You can also make up your own pizza with **extra** toppings. You can choose from four different extra toppings: olives, ham, mushrooms and salami.

Ross wants to order a pizza with two different **extra** toppings.

How many different combinations can Ross choose from?

Answer: .... combinations.

CHOICES – Question 1 illustrates Level 4 in PISA 2003 mathematics, with a difficulty of 559 PISA score points. On average across OECD countries, 49% of students were able to do this successfully.

Context: *Educational and occupational* – while the problem is located in an occupational setting, such a question is likely only to be found in a school mathematics classroom. Nevertheless the thinking involved is demanded in many situations and is clearly part of *mathematical literacy*.

Content area: *Quantity* – this problem belongs clearly to the field in mathematics called combinatorics. However, it is not necessary to use knowledge other than structured reasoning. From a mathematical point of view the problem is not too complex. However, from a reading point of view, it is. There is one pizza, with two basic ingredients, and four extra choices of which the student can choose two. A more or less structured and safe solution is to draw a basic pizza (represented here by B), and subsequently draw this pizza with all possibilities having one extra ingredient (and using the numbers 1 to 4 to represent the extras): B1, B2, B3 and B4. It is now possible to add the second extra ingredient that has to be different: (B11), B12, B13, B14, (B21), (B22), B23, B24, (B31), (B32), (B33), B34, (B41), (B42), (B43) and (B44). So, it is only possible to create 6 different pizzas: B12, B13, B14, B23, B24 and B34.

Competency cluster: *Connections* – students have clearly to mathematise the problem in the sense that they really have to read the text very precisely and identify the relevant information in a structured way. Next they have to come up with an answer that requires an organised and systematic way of thinking, making clear that all combinations have been found.

The question requires students to:

- Read and interpret a rather complex text.
- Identify the relevant mathematics.
- Develop a structured strategy to ensure finding all the answers.

Successful students answered "6". Students were most successful on this question in Japan (66%), Finland (60%), France and Korea (59%), the United Kingdom and Canada (58%).

On average across the OECD countries, most students attempted to answer this question – only 5% failed to respond.

EXPORTS

The graphics below show information about exports from Zedland, a country that uses zeds as its currency.

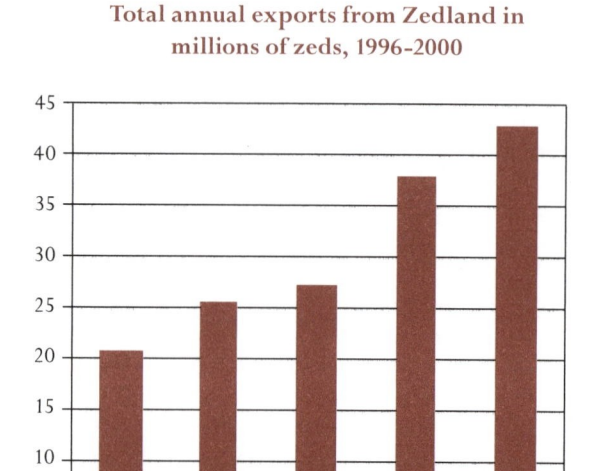

Total annual exports from Zedland in millions of zeds, 1996-2000

Distribution of exports from Zedland in 2000

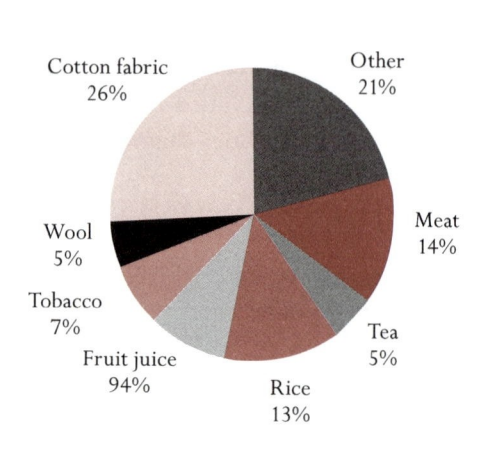

Question 2: EXPORTS

What was the value of fruit juice exported from Zedland in 2000?

  A   1.8 million zeds.

  B   2.3 million zeds.

  C   2.4 million zeds.

  D   3.4 million zeds.

  E   3.8 million zeds.

EXPORTS – Question 2 illustrates Level 4 in PISA 2003 mathematics, with a difficulty of 565 PISA score points. On average across OECD countries, 48% of students were able to do this successfully.

Context: *Public* – The information society in which we live relies heavily on data, and data are often represented in graphics. The media use graphics often to illustrate articles and make points more convincingly. Reading and understanding this kind of information therefore is an essential component of *mathematical literacy*.

Content area: *Uncertainty* – focus on using data. The mathematical content consists of reading data from two graphs: a bar chart and a pie chart, comparing the characteristics of the two graphs, and combining data from the two graphs in order to be able to carry out a basic number operation resulting in a numerical answer.

Competency cluster: *Connections* – combine the information of the two graphics in a relevant way. This mathematisation process has some distinct phases. Students need to decode the different standard representations by looking at the total of annual exports of 2000 (42.6) and at the percentage of the Fruit Juice exports (9%) of this total. Students then need to connect these numbers by an appropriate numerical operation (9% of 42.6).

The question requires students to:

- Use mathematical insight to connect and combine two graphical representations.
- Apply the appropriate basic mathematical routine in the relevant way.

Successful students chose answer "E" (3.8 million zeds). Students were most successful on this question in the partner economy Hong Kong-China (69%), the partner economy Macao-China (63%), the Netherlands (62%) and Belgium and the Czech Republic (60%). The most common incorrect answer chosen by students was "C" (16% on average across OECD countries), followed by "A" (11%) and "B" (10%). See Chapter 6 for additional discussion.

On average across the OECD countries 7% of students did not attempt to respond to this question, but this was the case for 16% of students in Italy and 20-21% of students in the partner countries Serbia and Uruguay.

ROBBERIES

Question 1: ROBBERIES

A TV reporter showed this graph and said:

"The graph shows that there is a huge increase in the number of robberies from 1998 to 1999."

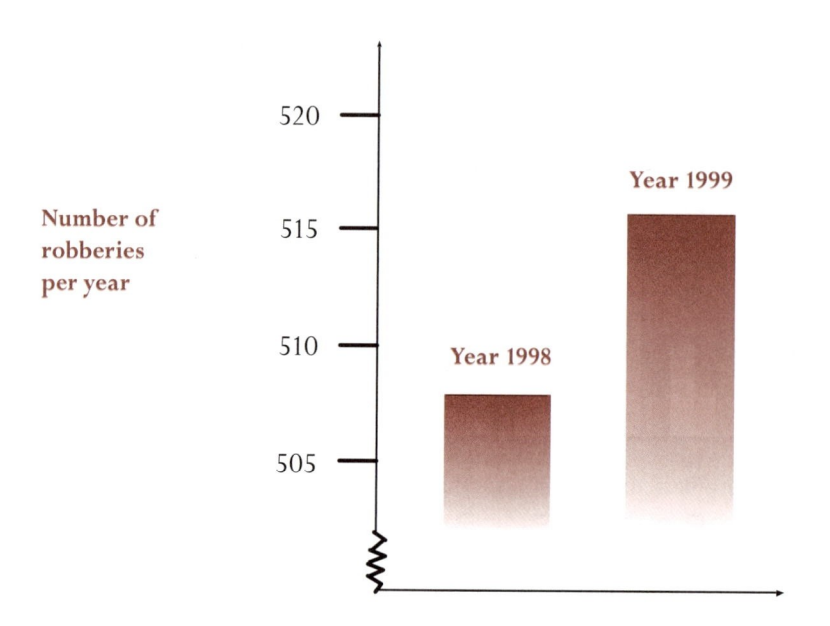

Number of robberies per year

Do you consider the reporter's statement to be a reasonable interpretation of the graph?

Give an explanation to support your answer.

ROBBERIES – Question 1 illustrates two levels of proficiency in PISA 2003 mathematics depending on whether students give partially or fully correct answers. Fully correct answers for this question illustrate Level 6, with a difficulty of 694 PISA score points. Here, a partially correct answer scored at 1 point illustrates performance at Level 4, with a difficulty of 577 PISA score points. On average across OECD countries, 28% of students were only capable of reaching this level of performance on ROBBERIES Q1.

Context: *Public* – The graph presented in this question was derived from a "real" graph with a similarly misleading message. The graph seems to indicate, as the TV reporter said: "a huge increase in the number of robberies". The students are asked if the statement fits the data It is very important to "look through" data and graphs as they are frequently presented in the media in order to function well in the knowledge society. This constitutes an essential skill in mathematical literacy. (See also the PISA Assessment Framework 2003, p. 105). Quite often designers of graphics use their skills (or lack thereof) to let the data support a pre-determined message, often with a political context. This is an example.

Content area: *Uncertainty* – analysis of a graph and interpretation of data. Understanding the issues related to misinterpretation of data. (In this graph the inappropriate cut in the y-axis indicates quite a large increase in the number of robberies, but the absolute difference between the number of robberies in 1998 and 1999 is far from dramatic).

Competency cluster: *Connections* – reasoning and interpretation competencies, together with communication skills.

The question requires students to:

- Understand and decode a graphical representation in a critical way.
- Make judgments and find appropriate argumentation based on mathematical thinking and reasoning (interpretation of data).
- Use some proportional reasoning in a statistical context and a non-familiar situation.
- Communicate effectively their reasoning process.

Students were considered partially correct when they indicated that the statement is not reasonable, but fail to explain their judgment in appropriate detail. Their reasoning only focuses on an increase given by an exact number of robberies in absolute terms, but not in relative terms. In some cases, students may communicate their answers ineffectively leaving their answers open to interpretation. For example: "an increase of around 10 is not large" could mean something different to "an increase from 508 to 515 is not large". The second answer shows the actual numbers, and thus could indicate that the increase is small due to the large numbers involved, but the first answer does not show this line of reasoning. Examples of partially correct answers include:

- Not reasonable. It increased by about 10 robberies. The word "huge" does not explain the reality of the increased number of robberies. The increase was only about 10 and I wouldn't call that "huge".
- From 508 to 515 is not a large increase.
- No, because 8 or 9 is not a large amount.
- Sort of. From 507 to 515 is an increase, but not huge.

Very few students in each country answered that the interpretation was not reasonable, but made an error in calculating the percentage increase. Such answers were also considered to be partially correct.

The following countries have the largest proportions of students who gave partially correct answers to this question: Finland (38%), Canada and Ireland (37%), the United Kingdom and Australia (36%) and Japan (35%).

Across the OECD countries on average, 15% of students did not respond to this question. This was the case for 30% of students in Greece, 28% in the Slovak Republic and 20% in Turkey, Mexico and Luxembourg, and for between 26 and 35% in the partner countries Serbia, Brazil, Uruguay, the Russian Federation, Tunisia and Indonesia.

GROWING UP

Youth grows taller

In 1998 the average height of both young males and young females in the Netherlands is represented in this graph

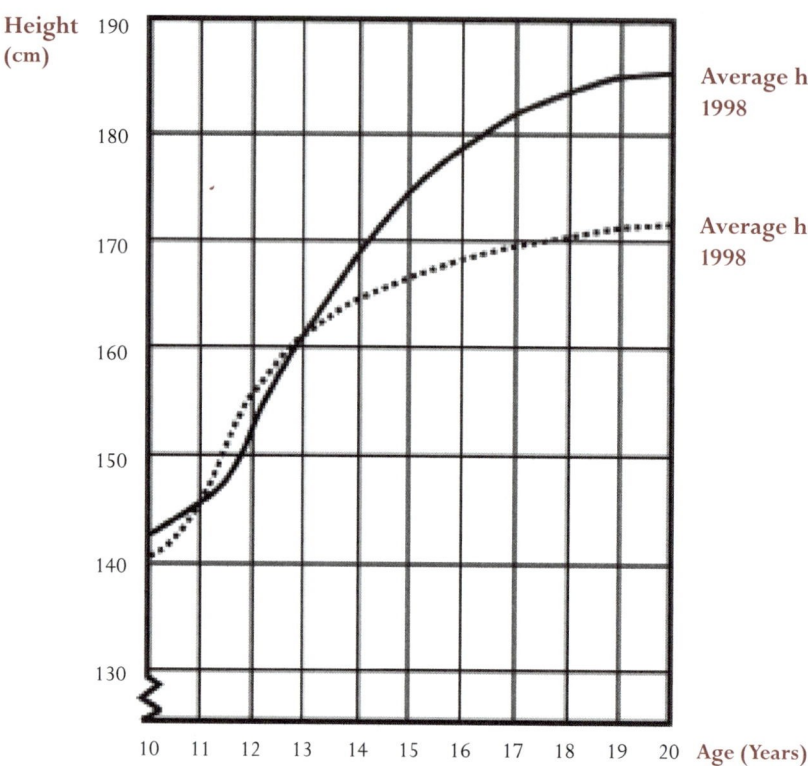

Height (cm)

Average height of young males 1998

Average height of young females 1998

Age (Years)

Question 3: GROWING UP

Explain how the graph shows that on average the growth rate for girls slows down after 12 years of age.

....
....
....

GROWING UP – Question 3 illustrates Level 4 in PISA 2003 mathematics with a difficulty of 574 PISA score points. On average across OECD countries, 45% of students were able to do this successfully.

Context: Scientific

Content area: Change and relationships – focus on the relationship between age and height. The mathematical concept of "decreasing growth". This is used often in the media, but seldom properly understood. The problem is the combination of "growing" and "slowing down", following the language used in the question.

In mathematical terms: the graphs should become less "steep". Even more mathematically: the slope (or gradient) would decrease.

Competency cluster: *Connections* – solve a problem in a non-routine situation, although still involving familiar settings. Students need to think and reason (what does the question mean in mathematical terms?), make an argument, and communicate this in a proper way (which is not trivial here). Students also need to solve the problem and decode the graph. The question is definitely not familiar and demands the intelligent linking of different ideas and information.

The question requires students to:

- Show mathematical insight.
- Analyse different growth curves.
- Evaluate the characteristics of a data set, represented in a graph.
- Note and interpret the different slopes at various points of the graphs.
- Reason and communicate the results of this process, within the explicit models of growth.

Successful students were able to read the graph correctly to determine that growth starts to diminish at age 12, or a bit before that age, and communicate this observation. Students were most successful on this question in the Netherlands (78%), Finland (68%) and Canada and Belgium (64%), where at least 88% of students responded to the question, compared to 79% on average across OECD countries. However, in some OECD countries significant proportions of students did not attempt to respond to this question, notably in Austria (44%) and Greece (43%).

In all countries successful students gave answers ranging from daily-life language to more mathematical language involving the reduced steepness, or they compared the actual growth in centimetres per year. Among the OECD countries, the most common correct answers were given in daily-life language. For example:

- It no longer goes straight up, it straightens out.
- The curve levels off.
- It is flatter after 12.
- The girls' line starts to even out and the boys' line just gets bigger.
- It straightens out and the boys' graph keeps rising.

This was the case for at least 70% of correct answers in 24 of the OECD countries, but only 39% in Korea and 49% in Austria. In Korea 56% of the correct answers were communicated in mathematical language, where students used terms such as "gradient", "slope", or "rate of change". This was the case for between 21 and 26% of correct answers in New Zealand, Turkey, Hungary, Canada, Japan and the Slovak Republic. In Austria, 34% of correct answers consisted of students comparing the actual growth. Examples of such answers include:

- From 10 to 12 the growth is about 15 cm, but from 12 to 20 the growth is only about 17 cm.
- The average growth rate from 10 to 12 is about 7.5 cm per year, but about 2 cm per year from 12 to 20 years.

Such answers comparing the actual growth also comprised a significant proportion of the correct answers in the following OECD countries: Mexico (26%), Greece (23%), France and Turkey (19%).

The most common error that students made was to give an answer that did not refer to the graph, for example "girls don't grow much after 12". However, around 40% of the incorrect answers given in France, Korea and Poland did refer to the graph, but simply indicated that the female height drops below the male height, without referring to the steepness of the female gradient.

EXCHANGE RATE

Mei-Ling from Singapore was preparing to go to South Africa for 3 months as an exchange student. She needed to change some Singapore dollars (SGD) into South African rand (ZAR).

Question 3: EXCHANGE RATE

During these 3 months the exchange rate had changed from 4.2 to 4.0 ZAR per SGD.

Was it in Mei-Ling's favour that the exchange rate now was 4.0 ZAR instead of 4.2 ZAR, when she changed her South African rand back to Singapore dollars? Give an explanation to support your answer.

EXCHANGE RATE – Question 3 illustrates Level 4 in PISA 2003 mathematics with a difficulty of 586 PISA score points. On average across OECD countries, 40% of students were able to do this successfully.

Context: *Public* – currency exchange associated with international travel

Content area: *Quantity* – quantitative relationships with money, procedural knowledge of number operations (multiplication and division)

Competency cluster: *Reflection* – students have to reflect on the concept of exchange rate and its consequences in this particular situation. This question illustrates the process of mathematisation. First students need to identify the relevant mathematics involved in this real-world problem. Although all the required information is explicitly presented in the question this is a somewhat complex task. Reducing the information in the question to a problem within the mathematical world places significant demands on students. Students need to think and reason flexibly (how do we find?), form an argument (how are the objects related?) and solve the mathematical problem. Combining these three competencies requires students to reflect on the process needed to solve the problem. Finally students need to communicate a real solution and explain the conclusion.

The question requires students to:

- Interpret a non-routine mathematical relationship (a specified change in the exchange rate for 1 Singapore Dollar/1 South African Rand).
- Reflect on this change.
- Use flexible reasoning to solve the problem.
- Apply some basic computational skills or quantitative comparison skills.
- Construct an explanation of their conclusion.

Students were most successful on this question in the partner country Liechtenstein (64%), Canada (58%), Belgium (55%), the partner economies Macao-China and Hong Kong-China (53%), Sweden, Finland and France (51%). Less than 20% of students answered this question correctly in Mexico and Turkey and in the partner countries Indonesia, Brazil, Thailand and Tunisia.

Forty-two percent of students responded to this question but gave a wrong answer, on average across OECD countries. In some cases, students answered "yes" but failed to give an adequate explanation or gave no explanation at all. For example:

- Yes, a lower exchange rate is better.
- Yes it was in Mei-Ling's favour, because if the ZAR goes down, then she will have more money to exchange into SGD.
- Yes it was in Mei-Ling's favour.

This was the case for 54% of the wrong answers in France and between 40% and 49% of the wrong answers in Ireland, New Zealand, Portugal, Switzerland, Australia, Austria, Greece, Finland, Spain, Luxembourg, Japan and the partner countries/economies the Russian Federation and Hong Kong-China.

A further 17% did not respond to the question and this was between 27 and 29% in Mexico, Italy, Portugal, Turkey, Greece, and the partner country Serbia, 35% in the partner country Tunisia and 42% in the partner country Brazil.

Note that this unit includes two other questions (EXCHANGE RATE – Question 1 and EXCHANGE RATE – Question 2) and these are presented in the section *Examples of easy questions in PISA 2003 mathematics*.

WALKING

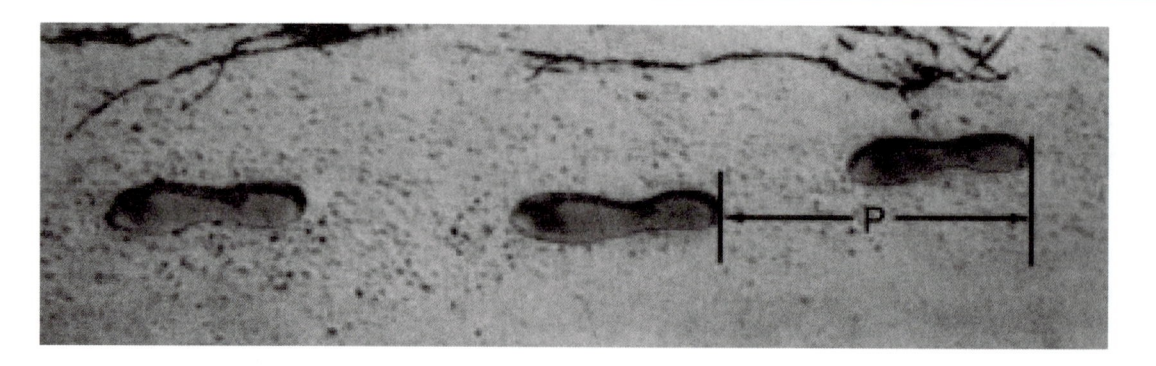

The picture shows the footprints of a man walking. The pacelength P is the distance between the rear of two consecutive footprints.

For men, the formula, $\frac{n}{P} = 140$, gives an approximate relationship between n and P where

n = number of steps per minute, and

P = pacelength in metres.

Question 3: WALKING

Bernard knows his pacelength is 0.80 metres. The formula applies to Bernard's walking.

Calculate Bernard's walking speed in metres per minute and in kilometres per hour. Show your working out.

WALKING – Question 3 illustrates three levels of proficiency in PISA 2003 mathematics depending on whether students give partially or fully correct answers. Fully correct answers for this item illustrate the high part of Level 6, with a difficulty of 723 PISA score points. There are two levels of partially correct answers: the higher level illustrates the higher part of Level 5, with a difficulty of 666 PISA score points (just 3 points below the boundary with Level 6) and the lower level illustrates the top part of Level 4, with a difficulty of 605 PISA score points (just 2 points below the boundary with Level 5). On average across OECD countries, 21% of students were able to do this successfully. Here we discuss the lower level of credit for WALKING Q3. This lower level received a score of 1 point and accounted for the 21% of responding students who were unable to do anything more on this problem.

Context: Personal

Content area: Change and relationships – the relationship between the number of steps per minute and pace-length. Conversion of measurement from m/min to km/hr.

The mathematical routine needed to solve the problem successfully is substitution in a simple formula (algebra), and carrying out a non-routine calculation. The first step in the solution process requires students to calculate the number of steps per minute when the pace-length is given (0.8 m). This requires proper substitution: $n/0.80 = 140$ and the observation that this equals: $n = 140 \times 0.80$ which in turn is 112 (steps per minute). The problem requires more than just routine operations: first substitution in an algebraic expression, followed by manipulating the resulting formula, in order to be able to carry out the required calculation. The next step is to go beyond the observation that the number of steps is 112. The question asks for the speed in m/minute: per minute he walks $112 \times 0.80 = 89.6$ meters; so his speed is 89.6 m/minute. The final step is to transform this speed from m/minute into km/h, which is a more commonly used unit of speed. This involves relationships among units for conversions within systems of units and for rates which is part of the measurement domain. Solving the problem also requires decoding and interpreting basic symbolic language in a less known situation, and handling expressions containing symbols and formulae.

Competency cluster: Connections – The problem is rather complex in the sense that not only is use of a formal algebraic expression required, but also doing a sequence of different but connected calculations that need proper understanding of transforming formulas and units of measures.

The question requires students to:

- Complete the conversions.
- Provide a correct answer in both of the requested units.

Students scoring at the lower level of partially correct answers includes those who wrote an expression that showed they had understood the formula and correctly substituted the appropriate values into it, finding the number of steps per minute. Such answers include:

- $n = 140 \times .80 = 112$. No further working out is shown or incorrect working out from this point.
- $n = 112$, 0.112 km/h.
- $n = 112$, 1120 km/h.
- 112 m/min, 504 km/h.

On average across OECD countries 20% of students were only able to achieve this lower level of partially correct answer. This was the case for 35% of students in the United States, 33% of students in Canada, 31% of students in the Slovak Republic and 30% of students in Greece.

EXAMPLES OF DIFFICULT MATHEMATICS QUESTIONS FROM PISA 2003

Twenty-seven mathematics questions from PISA 2003 lie in Levels 5 and 6 of the literacy scale. Nine of these relatively difficult questions have been released, and these are listed in Table 3.3 along with the difficulty of each question on the PISA mathematics scale, and other key framework characteristics. Note that two of these questions had one or more levels of partial credit associated with them, and the full and partial credit score points are listed in the table accordingly.

The absence of the *reproduction* competency cluster amongst the more difficult released questions seen in Table 3.3 is typical given the generally greater cognitive demands imposed by questions at Levels 5 and 6 of the PISA mathematics scale. All but one of the most difficult released questions are classified in either the *connections* or *reflection* competency clusters. The exception is a question requiring the application of routine knowledge and procedures, but using algebra in a real-world context. This is what makes the question more difficult than might otherwise be expected for questions in the *reproduction* competency cluster. The need to reflect substantively on the situation presented or on the solution obtained is a key challenge that tends to immediately make test questions more difficult than those for which such a demand is not made. The need to make connections among problem elements in order to solve a problem also makes questions more difficult compared to questions requiring the simple reproduction of practised knowledge and questions limited to the direct treatment of unconnected pieces of information.

Questions from all content areas, and from each of the context categories appear among the most difficult PISA mathematics questions, however only one of these is in the *quantity* area, and that question is not among the released set. Other more difficult *quantity* questions were developed for possible inclusion, but were not selected for the final PISA 2003 mathematics assessment.

Further, five of the nine questions require students to provide an extended response (either an extended sequence of calculations or an explanation or written argument in support of the conclusion). The most difficult questions typically have two features: the response structure is left open for the student, and active communication is required.

Finally, before introducing the more difficult released questions, the role of reading demand should be noted. Over half of the 27 most difficult questions in the PISA 2003 mathematics assessment are classified as "long", meaning they contain more than 100 words. The released questions listed in Table 3.3 also reflect this observation, since five of the nine questions are in this category. Reading demand is an important component of question difficulty.

Figures showing country level performances on many of these difficult items are found in Annex A1, Figures A.1.24 through A.1.31.

Table 3.3
Characteristics of the most difficult questions released from the PISA 2003 mathematics assessment

Item code	Question	OECD average percent correct	Location on PISA scale (PISA score points) Question	Location on PISA scale (PISA score points) Full/partial credit points	Traditional topic	Content area ("Overarching idea")	Competency cluster	Context ("Situation")	Length of question[1]	Response format[1]
M124Q01	WALKING – Question 1	36	611	611	Algebra	Change and relationships	Reproduction	Personal	Medium	Extended Response
M702Q01	SUPPORT FOR THE PRESIDENT – Question 1	36	615	615	Data	Uncertainty	Connections	Public	Long	Extended Response
M513Q01	TEST SCORES – Question 1	32	620	620	Data	Uncertainty	Connections	Educational and occupational	Long	Extended Response
M710Q01	FORECAST OF RAIN – Question 1	34	620	620	Data	Uncertainty	Connections	Public	Long	Multiple Choice
M402Q02	INTERNET RELAY CHAT – Question 2	29	636	636	Measurement	Change and relationships	Reflection	Personal	Long	Short Answer
M704Q02	THE BEST CAR – Question 2	25	657	657	Algebra	Change and Relationships	Reflection	Public	Long	Short Answer
M124Q03	WALKING – Question 3	21	665	605 666 723	Algebra	Change and relationships	Connections	Personal	Medium	Extended Response
M266Q01	CARPENTER – Question 1	20	687	687	Measurement	Space and shape	Connections	Educational and occupational	Medium	Complex Multiple Choice
M179Q01	ROBBERIES – Question 1[2]	30	635	577 694	Data	Uncertainty	Connections	Public	Short	Extended Response

1. Short questions contain fewer than 50 words. Medium-length questions contain 51 to 100 words. Long questions contain more than 100 words. Length of question in relation to question difficulty is discussed in detail in Chapter 5.

WALKING

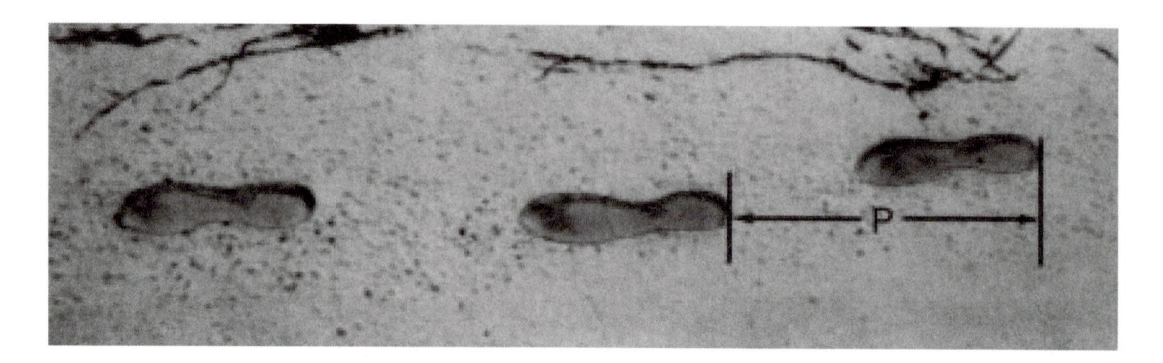

The picture shows the footprints of a man walking. The pacelength P is the distance between the rear of two consecutive footprints.

For men, the formula, $\dfrac{n}{P} = 140$, gives an approximate relationship between n and P where

  n = number of steps per minute, and

  P = pacelength in metres.

Question 1: WALKING

If the formula applies to Heiko's walking and Heiko takes 70 steps per minute, what is Heiko's pacelength? Show your work.

WALKING – Question 1 illustrates Level 5 in PISA 2003 mathematics, with a difficulty of 611 PISA score points (just four points beyond the boundary with Level 4). On average across OECD countries, 36% of students were able to do this successfully.

Context: *Personal* – Everyone has seen his or her own footsteps printed in the ground (whether in sand or mud) at some moment in life, most likely without realising the kind of relations that exist in the way these patterns are formed (although many students will have an intuitive feeling that if the pace-length increases, the number of steps per minute will decrease). To reflect on and realise the embedded mathematics in such daily phenomena is part of acquiring *mathematical literacy*.

Content area: *Change and relationships* – the relationship between the number of steps per minute and the pace-length. This relationship was derived from observing many different people walking steadily at their natural pace in a variety of situations. The mathematical content could be described as belonging clearly to algebra. Students need to solve the problem successfully by substituting in a simple formula and carrying out a routine calculation (70/p = 140) to find the value of p. The students need to carry out the actual calculation in order to get full credit.

Competency cluster: *Reproduction* – The competencies needed involve reproduction of practised knowledge, the performance of routine procedures, application of standard technical skills, manipulation of expressions containing symbols and formulae in standard form, and carrying out computations.

The question requires students to:

- Use a formal algebraic expression to solve a problem.

Successful students gave both the formula and the correct result. Examples of correct answers are:

$$0.5 \text{ m or } 50 \text{ cm, } \tfrac{1}{2} \text{ (unit not required)}.$$
$$70/p = 140$$
$$70 = 140\,p$$
$$p = 0.5$$
$$70/140$$

Students were most successful on this question in the partner economy Hong Kong-China (62%), the partner economy Macao-China (60%), the partner country the Russian Federation (54%), the Netherlands (52%) and the Slovak Republic (52%). On average across OECD countries 22% of students gave the correct formula but did not give the correct answer. This was the case for 48% of students in the United States, 35% of students in Ireland, 32% of students in Portugal and Luxembourg, 31% of students in Iceland, Poland and the partner country Indonesia. The original intention was to consider answers to be partially correct if students just gave the formula, but not the result or an incorrect result. However, the average ability of these students was not sufficiently higher than that of students who simply gave an incorrect answer. So no credit was awarded to students who only gave the formula.

Question 3: WALKING

Bernard knows his pacelength is 0.80 metres. The formula applies to Bernard's walking.

Calculate Bernard's walking speed in metres per minute and in kilometres per hour.

Show your work.

WALKING – Question 3 illustrates three levels of proficiency in PISA 2003 mathematics depending on whether students give partially or fully correct answers. Fully correct answers for this item illustrate the high part of Level 6, with a difficulty of 723 PISA score points. There are two levels of partially correct answers: the higher level illustrates the higher part of Level 5, with a difficulty of 666 PISA score points (just three points short of the boundary with Level 6) and the lower level discussed in the previous section dealing with Level 4, with a difficulty of 605 PISA score points. On average across OECD countries, 21% of students were able to solve this problem successfully.

Context: Personal

Content area: Change and relationships – the relationship between the number of steps per minute and pacelength. Conversion of measurement from m/min to km/hr.

The mathematical routine needed to solve the problem successfully is substitution in a simple formula (algebra), and carrying out a non-routine calculation. The first step in the solution process requires students to calculate the number of steps per minute when the pace-length is given (0.8 m). This requires proper substitution: $n/0.80 = 140$ and the observation that this equals: $n = 140 \times 0.80$ which in turn is 112 (steps per minute). The problem requires more than just routine operations: first substitution in an algebraic expression, followed by manipulating the resulting formula, in order to be able to carry out the required calculation. The next step is to go beyond the observation that the number of steps is 112. The question asks for the speed in m/minute: per minute he walks $112 \times 0.80 = 89.6$ meters; so his speed is 89.6 m/minute. The final step is to transform this speed from m/minute into km/h, which is a more commonly used unit of speed. This involves relationships among units for conversions within systems of units and for rates which is part of the measurement domain. Solving the problem also requires decoding and interpreting basic symbolic language in a less known situation, and handling expressions containing symbols and formulae.

Competency cluster: Connections – The problem is rather complex in the sense that not only is use of a formal algebraic expression required, but also doing a sequence of different but connected calculations that need proper understanding of transforming formulas and units of measures.

The question requires students to:

- Complete the conversions.
- Provide a correct answer in both of the requested units.

Successful students gave correct answers for both metres/minute and km/hour (although the units were not required). An example of a fully correct answer:

> $n = 140 \times .80 = 112$.
>
> Per minute he walks $112 \times .80$ metres = 89.6 metres.
>
> His speed is 89.6 metres per minute.
>
> So his speed is 5.38 or 5.4 km/hr.

Also fully correct answers do not need to show the working out (*e.g.* 89.6 and 5.4) and errors due to rounding are acceptable (*e.g.* 90 metres per minute and 5.3 km/hr [89 × 60]). However an answer of 5376 must specify the unit (m/hour) to be considered fully correct.

Students were most successful on this question in the partner economy Hong Kong-China (19% correct), Japan (18%), Belgium (16%), the Netherlands, the partner countries/economies Macao-China and Liechtenstein (15%) and Finland and Switzerland (14%).

Some students were able to find the number of steps per minute and make some progress towards converting this into the more standard units of speed asked for. However, their answers were not entirely complete or fully correct. Examples of partially correct answers include where:

- Students fail to multiply by 0.80 to convert from steps per minute to metres per minute. For example, his speed is 112 metres per minute and 6.72 km/hr.
- Students give the correct speed in metres per minute (89.6 metres per minute) but conversion to kilometres per hour is incorrect or missing.
- Students explicitly show the correct method, but make minor calculation error(s). No answers correct.

    $n = 140 \times .8 = 1120$; $1120 \times 0.8 = 896$. He walks 896 m/min, 53.76 km/h.

    $n = 140 \times .8 = 116$; $116 \times 0.8 = 92.8$. 92.8 m/min, 5.57 km/h.

- Students only give 5.4 km/hr, but not 89.6 metres/minute (intermediate calculations not shown).

On average across OECD countries 9% of students gave one of the above answers and were awarded the higher level of partially correct answers. This was the case for 30% of students in the partner economy Hong Kong-China, 26% in the partner economy Macao-China and 20% in Japan. Among these students the most common error was not to convert the number of steps into metres (this concerned around 70% of the higher level partially correct answers in Japan and the partner economies Hong Kong-China and Macao-China). Indeed, failure to convert the number of steps into metres was the most common reason why students just fell short of fully correct answers in 19 of the OECD countries. Conversely, the majority of students with higher level partially correct answers in Hungary, the Slovak Republic, Greece and Italy failed to convert from metres per minute into kilometres per hour (this concerned around 60% of such answers).

SUPPORT FOR THE PRESIDENT

Question 1: SUPPORT FOR THE PRESIDENT

In Zedland, opinion polls were conducted to find out the level of support for the President in the forthcoming election. Four newspaper publishers did separate nationwide polls. The results for the four newspaper polls are shown below:

Newspaper 1: 36.5% (poll conducted on January 6, with a sample of 500 randomly selected citizens with voting rights)

Newspaper 2: 41.0% (poll conducted on January 20, with a sample of 500 randomly selected citizens with voting rights)

Newspaper 3: 39.0% (poll conducted on January 20, with a sample of 1000 randomly selected citizens with voting rights)

Newspaper 4: 44.5% (poll conducted on January 20, with 1000 readers phoning in to vote).

Which newspaper's result is likely to be the best for predicting the level of support for the President if the election is held on January 25? Give two reasons to support your answer.

SUPPORT FOR THE PRESIDENT – Question 1 illustrates Level 5 in PISA 2003 mathematics, with a difficulty of 615 PISA score points. On average across OECD countries, 36% of students were able to do this successfully.

Context: Public – This problem illustrates an important aspect of mathematical literacy: the ability for citizens to critically judge presentations with a mathematical background. This is especially important for presentations like opinion polls that seem to be used increasingly in this media-centred society. Particularly when the articles or television items mention that the prediction or poll may not be "representative", or is not taken randomly, or is not "fair" in any other way. It is important not to simply accept such statements and results without looking closely at the data in the context of how they were collected.

Content area: Uncertainty – sampling. There are four important characteristics to evaluate the samples in the question: the more recent survey tends to be better, the survey should be taken from a large sample, it should be a random sample, and of course only respondents who are eligible to vote should be considered. To gain full credit students needed to come up with two of these four arguments, and therefore choose Newspaper 3.

Competency cluster: Connections – although some reflection may be helpful to the students. As well as needing a good understanding of sampling, students need to read a rather complex text and understand each of the four possibilities.

The question requires students to:

- Understand the text.
- Understand conceptually different aspects of sampling.
- Produce and write the reasons for choosing the answer given.
- Successful students answered "Newspaper 3" and gave at least two valid reasons to justify this conclusion. Possible reasons include: The poll is more recent, with larger sample size; a random selection of the sample; and only voters were asked. If students gave additional information (including irrelevant or

incorrect information) this was simply ignored. The essential was that they had given the correct answer with two valid reasons. Examples of correct answers include:

- Newspaper 3, because they have selected more citizens randomly with voting rights.
- Newspaper 3 because it has asked 1000 people, randomly selected, and the date is closer to the election date so the voters have less time to change their mind.
- Newspaper 3 because they were randomly selected and they had voting rights.
- Newspaper 3 because it surveyed more people closer to the date.
- Newspaper 3 because the 1000 people were randomly selected.

Students were most successful on this question in the partner economy Hong Kong-China (48%), France and Japan (47%), Finland, Canada, Australia, the Netherlands and Korea (46%) and New Zealand (45%). On average across OECD countries, 7% of students answered "Newspaper 3", but did not give an adequate explanation or gave no explanation. This was the case for less than 5% of students in Poland, Turkey, Japan and the partner economy Hong Kong-China.

TEST SCORES

Question 1: TEST SCORES

The diagram below shows the results on a Science test for two groups, labelled as Group A and Group B.

The mean score for Group A is 62.0 and the mean for Group B is 64.5. Students pass this test when their score is 50 or above.

Looking at the diagram, the teacher claims that Group B did better than Group A in this test.

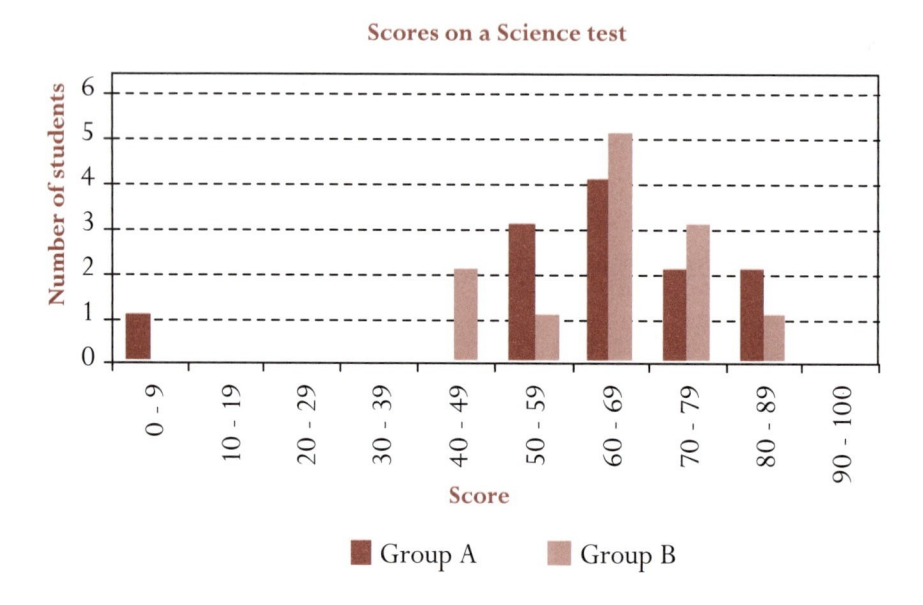

Scores on a Science test

[Bar chart showing Number of students (0-6) vs Score ranges (0-9, 10-19, 20-29, 30-39, 40-49, 50-59, 60-69, 70-79, 80-89, 90-100) for Group A and Group B.

Group A: 0-9: 1; 40-49: 0; 50-59: 3; 60-69: 4; 70-79: 2; 80-89: 2
Group B: 40-49: 2; 50-59: 1; 60-69: 5; 70-79: 3; 80-89: 1]

■ Group A   ■ Group B

The students in Group A don't agree with their teacher. They try to convince the teacher that Group B may not necessarily have done better.

Give one mathematical argument, using the graph, that the students in Group A could use.

TEST SCORES – Question 1 illustrates Level 5 in PISA 2003 mathematics, with a difficulty of 620 PISA score points. On average across OECD countries 32% of students were able to do this successfully.

Context: *Educational and occupational* – the educational context of this item is one that all students are familiar with: comparing test scores. In this case a science test has been administered to two groups of students: A and B. The results are given to the students in two different ways: in words with some data embedded and by means of two graphs in one grid.

Content area: *Uncertainty* – the field of exploratory data analysis. Knowledge of this area of mathematics is essential in the information society in which we live, as data and graphical representations play a major role in the media and in other aspects of daily experiences.

Competency cluster: *Connections* – includes competencies that not only build on those required for the *reproduction* competency cluster (like encoding and interpretation of simple graphical representations) but also require reasoning and insight, and in particular, mathematical argument. The problem is to find arguments that support the statement that Group A actually did better than group B, given the counter-argument of one teacher that group B did better – on the grounds of the higher mean for group B. Actually the students have a choice of at least three arguments here. The first one is that more students in group A pass the test; a second one is the distorting effect of the outlier in the results of group A; and finally Group A has more students that scored 80 or over. Another important competency needed is explaining matters that include relationships.

This question requires students to:

- Apply statistical knowledge in a problem situation that is somewhat structured and where the mathematical representation is partially apparent.
- Use reasoning and insight to interpret and analyse the given information.
- Communicate their reasons and arguments.

Many students did not respond to this question – 32% on average across OECD countries. Although this varies significantly among countries from 10% in the Netherlands and 13% in Canada to 49% in Mexico and the partner country Uruguay and 53% in Italy and 70% in the partner country Serbia.

Successful students gave one valid argument. Valid arguments could relate to the number of students passing, the disproportionate influence of the outlier, or the number of students with scores in the highest level. For example:

- More students in Group A than in Group B passed the test.
- If you ignore the weakest Group A student, the students in Group A do better than those in Group B.
- More Group A students than Group B students scored 80 or over.

Students were most successful on this question in the partner economies Hong Kong-China (64%) and Macao-China (55%) and in Japan (55%), Canada (47%), Korea (46%) and Belgium (44%).

On average across OECD countries, 33% of students responded to the question, but gave an incorrect answer. These included answers with no mathematical reasons, or wrong mathematical reasons, or answers that simply described differences but were not valid arguments that Group B may not have done better. For example:

- Group A students are normally better than Group B students in science. This test result is just a coincidence.
- Because the difference between the highest and lowest scores is smaller for Group B than for Group A.
- Group A has better score results in the 80-89 range and the 50-59 range.
- Group A has a larger inter-quartile range than Group B.

A significant proportion of students did not respond to this question (35% on average across OECD countries), although this varied from 11% in the Netherlands and 14% in Canada to 58% in Italy and Mexico, over 60% in the partner countries Tunisia and Uruguay and 73% in the partner country Serbia.

FORECAST OF RAINFALL

Question 1: FORECAST OF RAINFALL

On a particular day, the weather forecast predicts that from 12 noon to 6 pm the chance of rainfall is 30%. Which of the following statements is most likely to reflect the intended meaning of this forecast?

- A   30% of the land in the forecast area will get rain.
- B   30% of the 6 hours (a total of 108 minutes) will have rain.
- C   For the people in that area, 30 out of every 100 people will experience rain.
- D   If the same prediction was given for 100 days, then about 30 days out of the 100 days will have rain.
- E   The amount of rain will be 30% of a heavy rainfall (as measured by rainfall per unit time).

FORECAST OF RAINFALL – Question 1 illustrates a Level 5 item and has a difficulty level of 620 on the PISA score scale. On average across OECD countries, 34% of students were able to do this successfully.

Context: *Public* – forecast of rain is connected to media presentations and probability of events being held or cancelled

Content area: *Uncertainty* – involves the interpretation of factors and procedures associated with interpreting a statement involving probabilities

Competency cluster: *Connections* – students have to reflect on the concept of probability contained in a statement and use it to judge the validity of a number of statements

This question requires students to:

- Correctly interpret the given statement and connect it to the context described
- Use reflection and insight interpreting a standard probabilistic situation
- Compare and contrast the proposed communications based on information

Considerable variation was noted in student responses, ranging from a high of 54% correct in Korea and 49% correct in both Finland and partner country Liechtenstein to 11% in partner country Indonesia, 8% in Thailand, and 7% in partner country Tunisia.

INTERNET RELAY CHAT

Mark (from Sydney, Australia) and Hans (from Berlin, Germany) often communicate with each other using "chat" on the Internet. They have to log on to the Internet at the same time to be able to chat.

To find a suitable time to chat, Mark looked up a chart of world times and found the following:

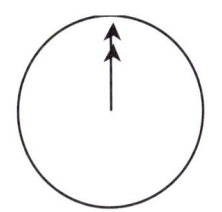

Greenwich 12 Midnight   Berlin 1:00 AM   Sydney 10:00 AM

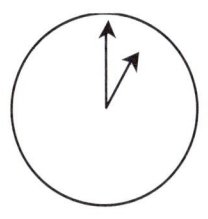

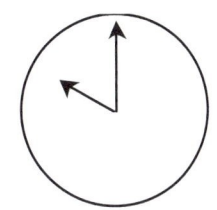

Question 2: INTERNET RELAY CHAT

Mark and Hans are not able to chat between 9:00 AM and 4:30 PM their local time, as they have to go to school. Also, from 11:00 PM till 7:00 AM their local time they won't be able to chat because they will be sleeping.

When would be a good time for Mark and Hans to chat? Write the local times in the table.

Place	Time
Sydney	
Berlin	

INTERNET RELAY CHAT – Question 2 illustrates Level 5 in PISA 2003 mathematics, with a difficulty of 636 PISA score points. On average across OECD countries, 29% were able to do this successfully.

Context: *Personal* – this assumes either that students are familiar at some level with chatting over the internet, and/or they know about time differences in this or another context. Given increasing globalisation and the enormous popularity of the internet this question really deals with *mathematical literacy*.

Content area: *Change and relationships* – time changes in different time zones.

Competency cluster: *Reflection* – rather high mathematisation skills are required to solve a non-routine problem. Students need to identify the relevant mathematics. Although the question seems rather straightforward, and the numbers and the actual mathematical operations required are rather simple, the question is actually more complex. The students have to understand the way that time spent sleeping and at school constrains the times that could be suitable for communicating with each other. First students need to identify the times that could work for each of them separately. Then, students have to compare two "time-windows" to find a time that would work for both of them simultaneously. This involves performing the same time calculation as in Question 1 of this unit, but within a context constrained by the students'

Learning Mathematics for Life: A Perspective from PISA – © OECD 2009

analysis of the possibilities. It is worth noting that the question could have been made more complex had the problem been to identify the whole window of opportunity. But the question requests the student to find just one particular time that would work, giving the students the opportunity to use trial-and-error methods.

The question requires students to:
- Understand the question.
- Mathematise the question.
- Identify one time that will work.

Successful students gave an answer with any time (*e.g.* Sydney 17:00, Berlin 8:00) or interval of time satisfying the 9 hours time difference. These could be taken from one of the following intervals:

Sydney: 4:30 PM – 6:00 PM; Berlin: 7:30 AM – 9:00 AM

Sydney: 7:00 AM – 8:00 AM; Berlin: 10:00 PM – 11:00 PM

If students gave an interval of time this needed to satisfy the constraints in its entirety. Also, students who did not specify morning (AM) or evening (PM), but gave times that could otherwise be regarded as correct, were given the benefit of the doubt and their answers were considered correct. Between 36% and 42% of students were successful on this question in New Zealand, Australia, Switzerland, Ireland, Canada, the Netherlands and Belgium, as well as in the partner country Liechtenstein. On average across OECD countries, 52% of students gave an incorrect answer (e.g. only one correct time) and 19% of students did not respond to the question. The highest percentages of students not responding to the question were in Denmark (31%), Spain (30%) and the partner country Serbia (45%).

THE BEST CAR

A car magazine uses a rating system to evaluate new cars, and gives the award of "The Car of the Year" to the car with the highest total score. Five new cars are being evaluated, and their ratings are shown in the table.

Car	Safety Features (S)	Fuel Efficiency (F)	External Appearance (E)	Internal Fittings (T)
Ca	3	1	2	3
M2	2	2	2	2
Sp	3	1	3	2
N1	1	3	3	3
KK	3	2	3	2

The ratings are interpreted as follows:

3 points = Excellent

2 points = Good

1 point = Fair

To calculate the total score for a car, the car magazine uses the following rule, which is a weighted sum of the individual score points:

$$\text{Total Score} = (3 \times S) + F + E + T$$

Question 2: THE BEST CAR

The manufacturer of car "Ca" thought the rule for the total score was unfair.

Write down a rule for calculating the total score so that Car "Ca" will be the winner.

Your rule should include all four of the variables, and you should write down your rule by filling in positive numbers in the four spaces in the equation below.

Total score = ……… × S + ……… × F + ……… × E + ……… × T

THE BEST CAR – Question 2 illustrates Level 5 in PISA 2003 mathematics with a difficulty of 657 PISA score points. On average across OECD countries, 25% of students were able to do this successfully.

Context: *Public* – an article in a car magazine is a very familiar context, especially for males. The underlying mathematics is relevant for males and females as everyone is presented with this kind of problem, that is, the evaluation of a consumer good using a rating system, whether it be cars, washing machines, coffee makers, etc. This is therefore an important part of *mathematical literacy*.

Content area: *Change and relationships* – the focus is on the relationship of numbers in a formula.

Competency cluster: *Reflection* – this is a complex problem as a whole and requires considerably advanced mathematical competencies. The question may not be very easy for students to understand. The idea that the car producer wants his car to win is rather simple. The complexity is that the new formula has to be valid for all cars, and still make "Ca" the winner. This involves considerable mathematical thinking and argumentation. Students need to identify the relevant mathematical concept of adding weights to different elements within a formula. In this case, students need to understand that the producer wants the strongest features of "Ca" (Safety and Interior) to be weighted most heavily. Plus, it is also desirable if the formula can minimise the stronger points of other cars, especially: External Appearance, and Fuel Efficiency. Using these arguments there are many possible correct answers. An example of a correct answer would be: (5V) + B + O + (5.I).

The question requires students to:

- Reflect on what the numbers in the formula really mean.
- Make the proper choices to weight the different elements within the formula correctly.
- Check the formula for correctness.

Successful students were able to provide a correct rule to make "Ca" the winner. Students were most successful on this question in Japan (45% correct), the partner economy Hong Kong-China (40%), Korea (38%), Belgium (37%) and Switzerland (36%). In seven OECD countries only 20% or fewer students were able to do this successfully.

On average across the OECD countries, 19% of students did not respond to this question, but this concerned 32% of students in Denmark and 31% of students in Italy.

Note that this unit includes one other question (THE BEST CAR – Question 1) and this is presented in the section *Examples of easy questions in PISA 2003 mathematics*.

CARPENTER

Question 1: CARPENTER

A carpenter has 32 metres of timber and wants to make a border around a garden bed. He is considering the following designs for the garden bed.

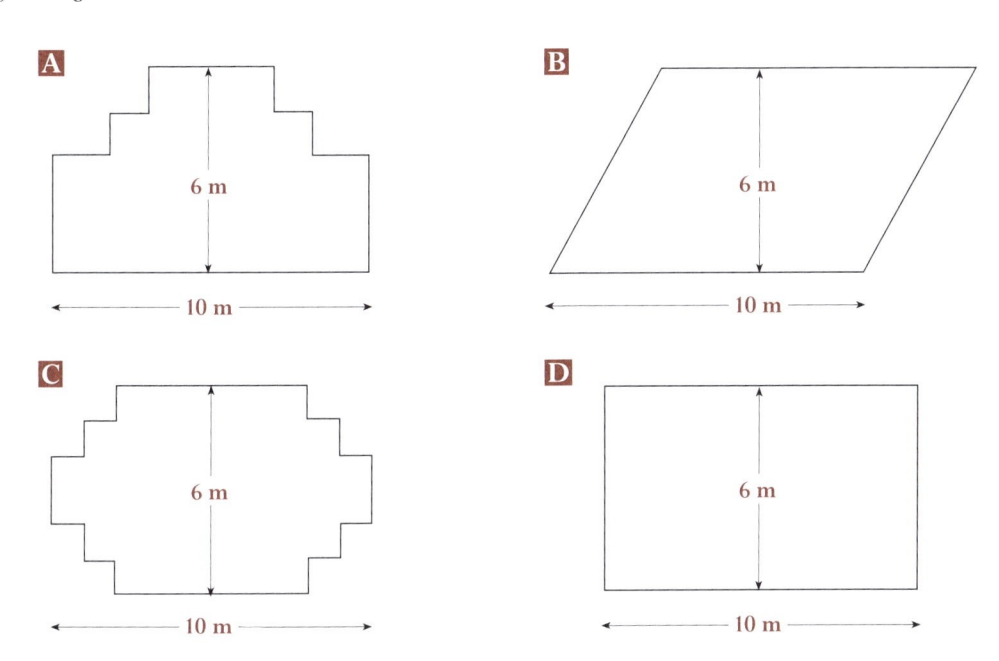

A

6 m

10 m

B

6 m

10 m

C

6 m

10 m

D

6 m

10 m

Circle either "Yes" or "No" for each design to indicate whether the garden bed can be made with 32 metres of timber.

Garden bed design	Using this design, can the garden bed be made with 32 metres of timber?
Design A	Yes / No
Design B	Yes / No
Design C	Yes / No
Design D	Yes / No

CARPENTER – Question 1 illustrates Level 6, with a difficulty of 687 PISA score points. On average across OECD countries, 20% of students were able to do this successfully.

Context: *Educational and occupational* – it is the kind of "quasi-realistic" problem that would typically be seen in a mathematics class, rather than being a genuine problem likely to be met in an occupational setting. Whilst not regarded as typical, a small number of such problems have been included in the PISA assessment. However, the competencies needed for this problem are certainly relevant and part of *mathematical literacy*.

Content area: *Space and shape* – geometrical knowledge.

Competency cluster: *Connections* – the problem is certainly non-routine. The students need the competence to recognise that for the purpose of solving the question the two-dimensional shapes A, C and D have the same perimeter; therefore they need to decode the visual information and see similarities and differences. The students need to see whether or not a certain border-shape can be made with 32 metres of timber. In three cases this is rather evident because of the rectangular shapes. But the fourth (B) is a parallelogram, requiring more than 32 metres.

The question requires students to:

- Decode visual information.
- Use argumentation skills.
- Use some technical geometrical knowledge and geometrical insight.
- Use sustained logical thinking.

Successful students answered "Design A, Yes; Design B, No; Design C, Yes; Design D, Yes". Students were most successful on this question in the partner economy Hong Kong-China (40% correct), Japan (38%), Korea (35%) and the partner economy Macao-China (33%). Less than 10% of students were able to do this successfully in Mexico, Greece and the partner countries Tunisia and Brazil. Nearly all students attempted to answer this question with only 2% failing to do so, on average across OECD countries, and this non-response rate did not surpass 5% of students in any of the OECD countries.

Note: There are actually four questions that students need to answer and this format is often associated with higher question difficulty, since students have to provide the correct response to all parts of the question in order to give a fully correct answer. The sustained logical thinking required to answer all question parts typically indicates a strong understanding of the underlying mathematical issues. On average across OECD countries, 31% of students gave three out of four correct answers. This ranged from 24% of students in Mexico and Turkey to 36% of students in Finland and Denmark. The majority of students across OECD countries tried to answer the question (on average only 2% failed to do so). However, several students had limited success in this. In fact 26% of students on average across OECD countries only gave one out of four correct answers. This was the case for at least 30% of students in Mexico, Greece, Turkey, the United States, Ireland, Portugal and Spain.

ROBBERIES

Question 1: ROBBERIES

A TV reporter showed this graph and said:

"The graph shows that there is a huge increase in the number of robberies from 1998 to 1999."

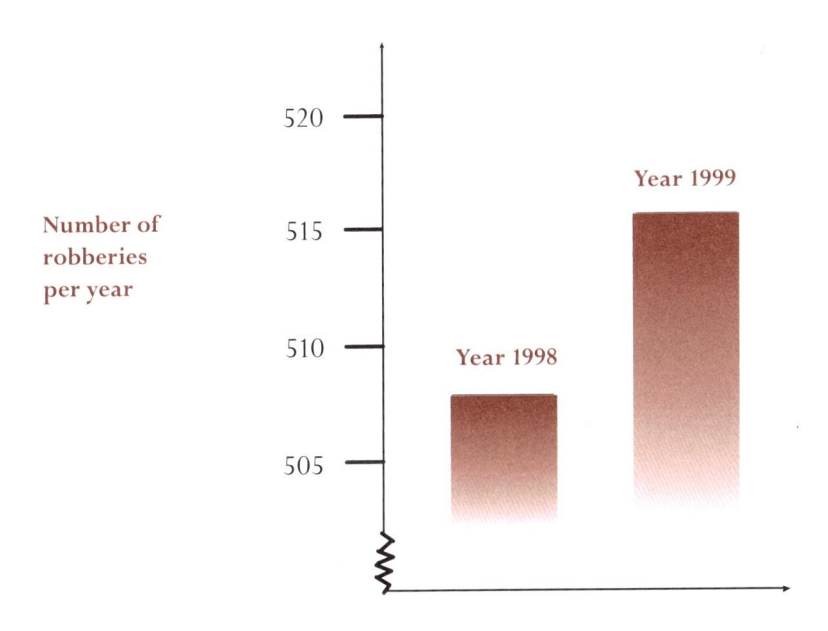

Do you consider the reporter's statement to be a reasonable interpretation of the graph?

Give an explanation to support your answer.

ROBBERIES – Question 1 illustrates two levels of proficiency in PISA 2003 mathematics depending on whether students give partially or fully correct answers. The latter were discussed in the previous section. Fully correct answers for this question illustrate Level 6, with a difficulty of 694 PISA score points. On average across OECD countries, 30% of students were able to do this successfully.

Context: *Public* – The graph presented in this question was derived from a "real" graph with a similarly misleading message. The graph seems to indicate, as the TV reporter said: "a huge increase in the number of robberies". The students are asked if the statement fits the data It is very important to "look through" data and graphs as they are frequently presented in the media in order to function well in the knowledge society. This constitutes an essential skill in mathematical literacy. (See also the *PISA Assessment Framework 2003*, p. 105). Quite often designers of graphics use their skills (or lack thereof) to let the data support a pre-determined message, often with a political context. This is an example.

Content area: *Uncertainty* – analysis of a graph and interpretation of data. Understanding the issues related to misinterpretation of data. (In this graph the inappropriate cut in the y-axis indicates quite a large increase in the number of robberies, but the absolute difference between the number of robberies in 1998 and 1999 is far from dramatic).

Competency cluster: *Connections* – reasoning and interpretation competencies, together with communication skills.

The question requires students to:

- Understand and decode a graphical representation in a critical way.
- Make judgments and find appropriate argumentation based on mathematical thinking and reasoning (interpretation of data).
- Use some proportional reasoning in a statistical context and a non-familiar situation.
- Communicate effectively their reasoning process.

Successful students indicate that the statement is not reasonable, and explain their judgment in appropriate detail. Their reasoning focuses on the increase of robberies in relative terms and not only on the increase given by an exact number of robberies in absolute terms. Students were most successful on this question giving fully correct answers in Sweden (32% correct), Norway (29%), Finland (27%), Belgium (24%), Italy, New Zealand, Canada and the partner economy Hong Kong-China (23%), Australia and the Netherlands (22%).

Among the OECD countries, the most common type of fully correct answers given by students comprised arguments that the entire graph should be displayed. For example:

- *I don't think it is a reasonable interpretation of the graph because if they were to show the whole graph you would see that there is only a slight increase in robberies.*
- *No, because he has used the top bit of the graph and if you looked at the whole graph from 0 to 520, it wouldn't have risen so much.*
- *No, because the graph makes it look like there's been a big increase but you look at the numbers and there's not much of an increase.*

Such arguments represented at least 70% of the correct answers given in Norway, New Zealand, the United States, Spain, Canada and the United Kingdom.

A significant proportion of fully correct answers given by students also included arguments in terms of the ratio or percentage increase. For example:

- *No, not reasonable. ten is not a huge increase compared to a total of 500.*
- *No, not reasonable. According to the percentage, the increase is only about 2%.*
- *No. eight more robberies is 1.5% increase. Not much in my opinion!*
- *No, only eight or nine more for this year. Compared to 507, it is not a large number.*

Such arguments represented at least 50% of the correct answers given in Japan, the Czech Republic, Turkey, Italy and Greece, and between 40 and 49% in Austria, France, the Slovak Republic, Switzerland, Portugal, Germany, Poland, Denmark and Ireland.

A minority of the fully correct answers given by students included arguments that trend data are required in order to make such a judgment. For example:

- *We cannot tell whether the increase is huge or not. If in 1997, the number of robberies is the same as in 1998, then we could say there is a huge increase in 1999.*
- *There is no way of knowing what "huge" is because you need at least two changes to think one huge and one small.*

While such arguments represented less than 10% of fully correct arguments given in most OECD countries, this was the case for 32% in Korea, 20% in Mexico and 16% in Japan and the Slovak Republic.

Across the OECD countries on average, 15% of students did not respond to this question. This was the case for 30% of students in Greece, 28% in the Slovak Republic and 20% in Turkey, Mexico and Luxembourg, and for between 26 and 35% in the partner countries Serbia, Brazil, Uruguay, the Russian Federation, Tunisia and Indonesia.

CONCLUSION

This chapter illustrated and discussed units and questions of varying difficulties and analysing them in relationship to characteristics that are likely to contribute to making them more or less difficult.

Aspects of the context the question is presented are important in this regard. First, it can be conjectured that contexts that are artificial and that play no role in solving a problem are likely to be less engaging than contexts that both hold more intrinsic interest and are critical to understanding the problem and its solution. On the other hand, contextualised problems that require students to make connections between the problem context and the mathematics needed to solve the problem place a different kind of demand on students. This kind of demand is frequently observed in only the most difficult questions.

Aspects of the question format and particularly of the response requirements are also very important determinants of question difficulty. Questions requiring students to select a response from a number of given options tend to be easier, but this is not always the case, particularly where students must do this a number of times within a single question, *i.e,.* for questions with the complex multiple-choice format. In those questions, a degree of sustained thought is required that exposes the thoroughness of the students' understanding of the mathematical concepts and skills involved in solving the problem.

Questions providing clear direction as to the nature of the answer required, and where convergent thinking is called for to find the one answer that is possible, are usually relatively easy. At the opposite end of the spectrum are questions that require students to construct a response with little or no guidance as to what would constitute an acceptable answer, and where a number of different answers might be acceptable. These questions tend to be more difficult than questions having a more convergent and closed format. When there is an added expectation for students to write an explanation of their conclusion or a justification of their result questions can become very difficult indeed.

Questions with a greater reading load also tend to be more difficult. Sometimes this may be influenced by the extra effort required when more words are involved, but the specific language elements used can also contribute to the level of difficulty. More technical words are less readily handled than simpler words.

Chapters 4 and 5 examine in more detail how students perform on these different types of questions, by analysing their performance on the complete PISA 2003 mathematics question set.

Comparison of Country Level Results

This chapter focuses on differences in the patterns of student performance by aspects of mathematical content contained within PISA 2003 assessment items' expectations. In participating countries, by the age of 15, students have been taught different subtopics from the broad mathematics curriculum. The subtopics vary in how they are presented to the students depending on the instructional traditions of the country.

INTRODUCTION

The PISA 2003 assessment framework (OECD, 2003) emphasises that "*mathematical literacy* focuses on the capacity of 15-year-olds (the age when many students are completing their formal compulsory mathematics learning) to use their mathematical knowledge and understanding to help make sense of [issues affecting engaged citizens in their real-worlds] and to carry out the resulting tasks" (p. 24). However, the amount and content of this knowledge 15-year-olds hold is largely dependent on what they have learnt at school. This learning appears to vary greatly across schools between and within countries.

Differences in curricula and traditions …

Evidence from the Third International Mathematics Study (TIMSS) that "the countries' traditions in mathematics education placed unequal emphasis on these subtopics in the curriculum, and as a consequence of this the students' performances were also quite different" (Zabulionis, 2001). Similar results regarding patterns of performance were obtained by Wu (2006) in relation to PISA 2000. That is, countries with similar mathematics curriculum also had similar response patterns in assessments of student capabilities within mathematics.

The substantive analysis for this chapter begins with an investigation of the patterns of performance by country. Then, the relative difficulty of particular topics and individual items is examined after an adjustment that places each country's overall mean difficulty to 0 in order that comparisons can be made among countries.

… and in grade level partly explain performance patterns across countries …

The report continues with grade level differences in performance. Due to the differences in grade level, the knowledge accumulated by age 15 can be quite different even in the same country. This also can influence students' performance. Previous research related to TIMSS indicated that there were significant differences between countries in some topics depending on whether the topic had been taught or not (Routitsky and Zammit, 2002).

Similar results can be found in PISA depending on the country and year level of students. This chapter examines a breakdown of students' performance by country, mathematics topic and grade level to investigate the impact of curriculum and instructional traditions on the patterns of performance.

… competency clusters and context areas.

The chapter concludes with examination of item difficulty by competency clusters and context areas by country and overall. While all competency clusters are important for *mathematical literacy*, it is equally important to balance instruction in terms of difficulty. There is a wide-spread belief that some contexts are more relevant for students than others, PISA provides rich data for examination of this subject.

CROSS COUNTRY DIFFERENCES IN CURRICULUM

Using the TIMSS 1996 test item data, Zabulionis (2001) categorised participating countries' achievement patterns into four groups, using a hierarchical cluster analysis. The four groups that resulted were characterised as follows:

- English-Speaking Group: Australia, Canada, England, Ireland, New Zealand, Scotland and the United States.

- Post-Communist Group: Bulgaria, Czech Republic, Hungary, Latvia and Lithuania.
The Russian Federation, Romania, Slovak Republic and Slovenia.

- Nordic Group: Denmark, Iceland, Norway and Sweden.

- Eastern Asian Group: Hong Kong-China, Japan, Korea, and Singapore.

Previous research identifies four groups of countries with similar performance.

This chapter attempts to link country performance, at the individual item level or for groups of items, to mathematics instruction in the countries participating in PISA 2003. By examining relative differences between countries in performance related to recognisable subtopics of mathematics, one can identify potential *mathematical literacy* weaknesses and strengths within each country. This information can provide valuable feedback for curriculum design and instructional practices. It should be noted that the comparisons carried out in this chapter focus on relative differences in performance in subtopics of mathematics within each country. That is, regardless of how well a country performed overall, relative weaknesses and strengths are identified within each country. In this way, the comparisons across countries are not simply based on a horse-race ranking of countries. Rather, the comparisons use yardsticks within each country for reference. For high performing countries, there may still be room for improvement in striking a balance in curriculum design. For low performing countries, specific areas of mathematics may be identified as trouble spots. In this way, the PISA survey can provide information beyond a simple ranking of countries, and, in doing so, relate PISA 2003 findings to potential improvements in instructional practices unique to the countries.

In addition to identifying differential performance across countries, differential performance at the item level between adjacent grades within each country may also reveal defining features or deficits in their mathematics curriculum structure. Previous research related to TIMSS indicated that there were significant differences between countries in some topics depending on whether the topic had been taught or not taught (Routitsky and Zammit, 2002). Such differences, if found in the PISA survey, could provide insight into the relationship of instruction to student performance. For example, if a country's curriculum has a spiral structure (mathematics topics are repeated from one grade to another, but with further extensions at a higher grade level), and then one might expect smaller differences between performances across grades, at the item level. But if the curriculum has a linear structure (different topics are taught at different grade levels), then one might expect greater differences in performance across different grades, at the item level, especially if little use of the content is made in the grade level in which students are assessed.

GROUPINGS OF COUNTRIES BY PATTERNS IN ITEM RESPONSES

To identify the extent to which groups of countries have similar patterns of item responses, an analysis was carried out to obtain the relative difficulties of PISA 2003 mathematics items within each country. For example, if two countries have similar curricula, one would expect the relative item difficulties to be similar for these two countries. So, if item A is more difficult than item B for Country 1, then item A is expected to be more difficult than item B for Country 2, even if the overall performance of Country 1 is a great deal higher (or lower) than that of Country 2. Such a comparison of relative item difficulty can be carried out by comparing separately calibrated item parameters (*e.g.* difficulty, discrimination, guessing) for each country, where the mean of the item difficulties for each country is set to zero. In this way, comparisons of relative item difficulties can be made between countries, without being confounded by the overall ability of the students in each country. When an item appears to be more (or less) difficult for students in some countries than for students in other countries, it is said that the item exhibits differential item functioning (DIF) with respect to the variable "country".

Performance patterns in PISA questions can provide useful insights for curriculum design.

In general, one would expect difficult items to be difficult for most countries, and easy items to be easy for most countries. Figure 4.1 shows a plot of item parameters by country, for three selected items. It can be seen that, overall, a difficult item (*e.g.* THE THERMOMETER CRICKET Q2) is difficult for all countries, and an easy item (*e.g.* A VIEW WITH A ROOM Q1) is easy for all countries. BRICKS Q1 was slightly above average difficulty across all of the countries. However, for each item, there are small variations among countries. These variations are expected, given that there are differences between countries in language, culture, curriculum structure, teaching methodology, and many other factors. But what is interesting is that there do appear to be patterns of groupings of countries that exhibit the same variation in the item parameters. For example, Hong Kong-China, Macao-China and Korea all found THE THERMOMETER CRICKET Q2 relatively easier as compared to other countries. In contrast, Brazil and Portugal found this item more difficult as compared to other countries. The question THE THERMOMETER CRICKET Q2 requires students to understand relationships between variables and to express the relationships algebraically.

It is interesting to study each item and the skills required for the item, and note each country's relative performance on the item, and the extent to which the item exhibits DIF. Unfortunately, such a study will not be overly useful in informing mathematics teaching in general, as the results relate only to isolated items. However, on the other hand, analyses of groups of countries with similar patterns of calibrated item parameters have the potential to provide more powerful information on which to form hypotheses about curriculum structures within groups of countries.

To identify groups of countries with similar patterns of performance, a hierarchical cluster analysis[1] was carried out on the separately calibrated item parameters for countries and sub-regions. Sub-regions with different languages are included in the cluster analysis to provide some information on the importance of language on student performance results. Figure 4.2 shows the dendrogram generated by the cluster analysis. This diagram shows from bottom to top the order in which similar countries join together in "shortest distances" between the joining countries in terms of the patterns of item difficulty parameters. For example, Australia and New Zealand are the two closest countries in terms of their patterns of item difficulties. They are then joined by the United Kingdom, and then joined by Canada (English) and Scotland.

Figure 4.1 ■ **Comparison of item parameters by countries for three selected items**

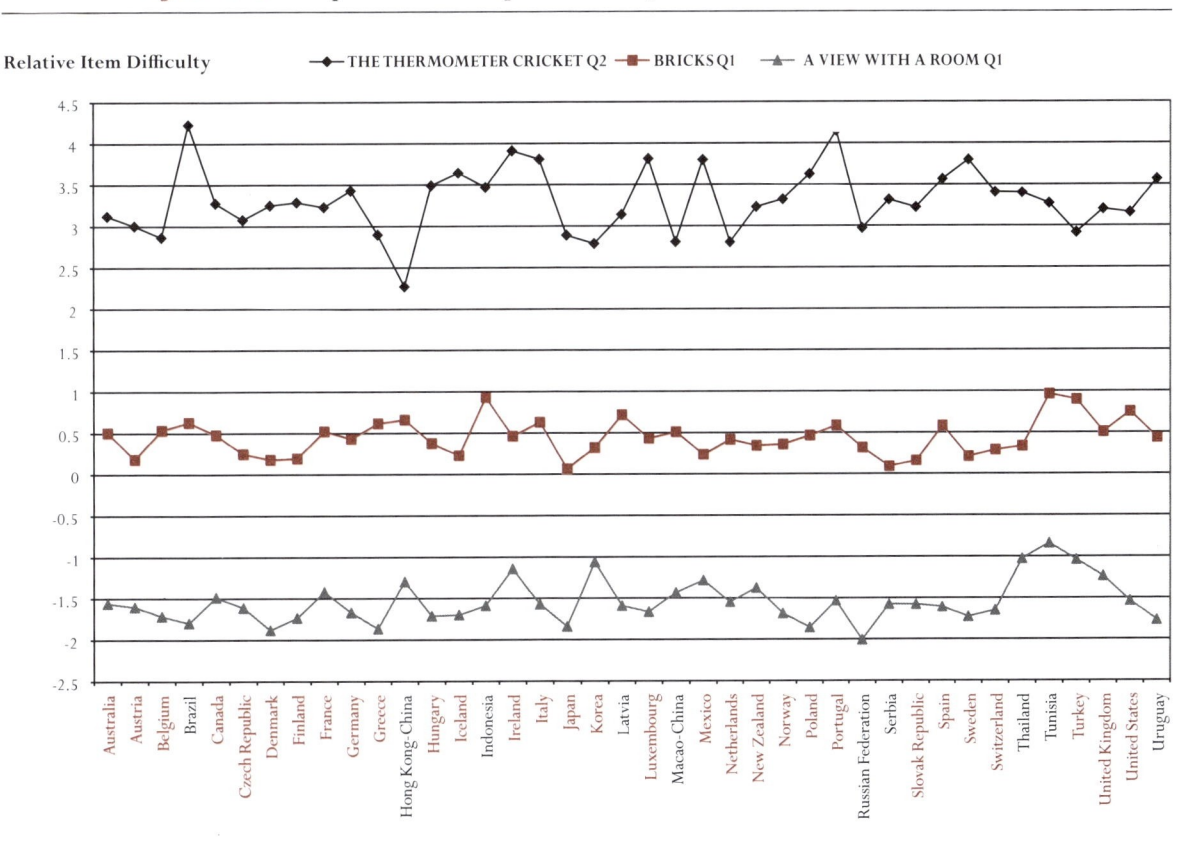

Performance in PISA suggests the following two large groups of countries:
i) English speaking countries (except the United States) and Scandinavian countries

ii) European countries.

The results from the cluster analysis can be summarised as follows (and also shown by visually grouped countries in Figure 4.2). Apart from the United States, English-speaking countries form a cluster grouping with similar performance patterns. They are joined by Scandinavian countries, however the Scandinavian countries (Finland, Denmark, Norway and Sweden) are more similar to the English-speaking countries than they are to each other. Further, the Scandinavian countries are closer to English-speaking countries than to other European countries.

European countries form a cluster, and in particular, countries sharing the same language tend to have very similar performance patterns. For example, Germany and the German-speaking part of Luxembourg, Italy and the Italian-speaking part of Switzerland, Austria and the German-speaking part of Switzerland, etc., all show close links with each other in their patterns of item difficulties. Interestingly, Eastern European countries such as the Czech Republic, the Slovak Republic, Hungary, Poland and the Latvian-speaking part of Latvia all show closer links to Western European countries than to the Russian Federation, a result that is different from the findings of earlier studies such as the one carried out by Zabulionis (2001).

East Asian countries such as Japan, Hong Kong-China and Korea are somewhat different from each other, as well as different from English or European groups. Note, however, that Hong Kong-China and Macao-China are very closely linked, and Japan and Korea are closer together than they are to other countries.

It is difficult to clearly identify the factors accounting for the observed clusters, since language, culture, geographical locations and educational traditions are so intertwined that it is nearly impossible to clearly separate the four. Some may suggest that the clusters are simply formed by language groups. This is not quite right. For example, the French-speaking part of Canada is closer to the English-speaking countries and Brazil is closer to Mexico than to Portugal.

While the underlying reasons for the observed clustering may be difficult to identify, the results of the cluster analysis provide us with a starting point for making further hypotheses and investigation.

PATTERNS IN MATHEMATICS CONTENT

PISA questions can be classified by traditional mathematic domains.

Given that some countries found particular items more (or less) difficult than other countries, it would be interesting to examine whether there exist patterns in the DIF results for each mathematics topic. For example, a country may perform consistently better (or worse) on a particular mathematics topic relative to other topics. Or, perhaps, groups of countries may show the same pattern across different mathematics topics, depending on the way mathematics is taught in the countries.

For such an analysis to be carried out, PISA mathematics items first need to be classified according to traditional curriculum topics. Analyses can then be carried out for specific mathematics topics and for specific groups of countries. In this study,

Figure 4.2 ■ **Hierarchical cluster analysis of item parameters**

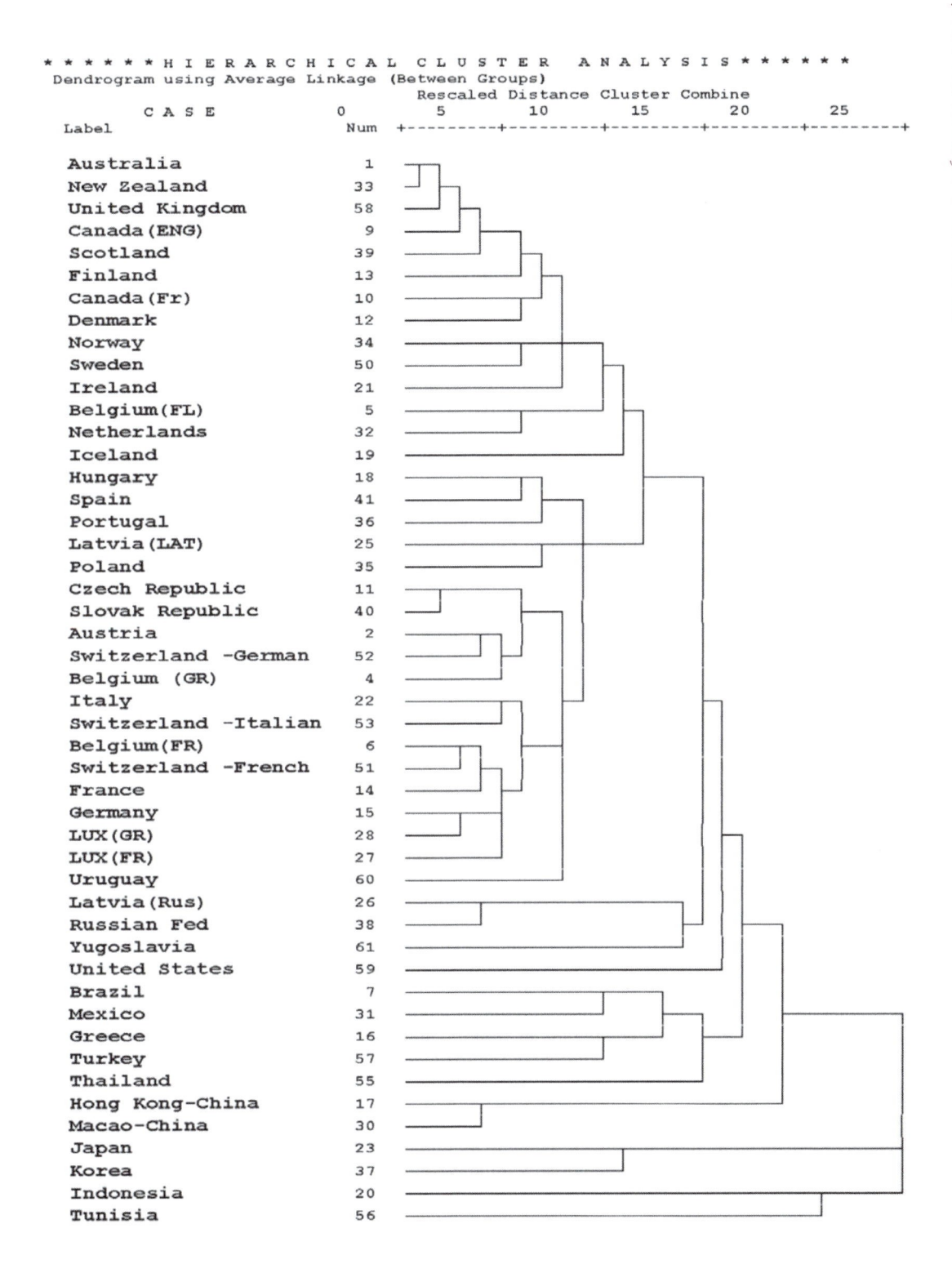

```
* * * * * * H I E R A R C H I C A L   C L U S T E R   A N A L Y S I S * * * * * *
Dendrogram using Average Linkage (Between Groups)
                              Rescaled Distance Cluster Combine
         C A S E          0        5       10       15       20       25
         Label           Num   +--------+--------+--------+--------+--------+

    Australia              1    ─┐
    New Zealand           33    ─┤
    United Kingdom        58    ─┤
    Canada(ENG)            9    ─┤
    Scotland              39    ─┤
    Finland               13    ─┤
    Canada(Fr)            10    ─┤
    Denmark               12    ─┤
    Norway                34    ─┤
    Sweden                50    ─┤
    Ireland               21    ─┤
    Belgium(FL)            5    ─┤
    Netherlands           32    ─┤
    Iceland               19    ─┤
    Hungary               18    ─┤
    Spain                 41    ─┤
    Portugal              36    ─┤
    Latvia(LAT)           25    ─┤
    Poland                35    ─┤
    Czech Republic        11    ─┤
    Slovak Republic       40    ─┤
    Austria                2    ─┤
    Switzerland -German   52    ─┤
    Belgium (GR)           4    ─┤
    Italy                 22    ─┤
    Switzerland -Italian  53    ─┤
    Belgium(FR)            6    ─┤
    Switzerland -French   51    ─┤
    France                14    ─┤
    Germany               15    ─┤
    LUX(GR)               28    ─┤
    LUX(FR)               27    ─┤
    Uruguay               60    ─┤
    Latvia(Rus)           26    ─┤
    Russian Fed           38    ─┤
    Yugoslavia            61    ─┤
    United States         59    ─┤
    Brazil                 7    ─┤
    Mexico                31    ─┤
    Greece                16    ─┤
    Turkey                57    ─┤
    Thailand              55    ─┤
    Hong Kong-China       17    ─┤
    Macao-China           30    ─┤
    Japan                 23    ─┤
    Korea                 37    ─┤
    Indonesia             20    ─┤
    Tunisia               56    ─┘
```

the PISA mathematics items have been classified under the following five general mathematics curriculum topics: Number, Algebra, Measurement, Geometry and Data. These five curriculum topics are typically included in national curriculum documents in many countries. They also match the TIMSS framework classifications. Consequently, the information collected in the Test-Curriculum Match Analysis (TCMA) in TIMSS (see Chapter 5 of *TIMSS 2003: International Mathematics Report* [Mullis, Martin, Gonzalez, and Chrostowski, 2004]) can also be used as supporting evidence to link curriculum to test results.

For each country, item difficulties for items classified under the same mathematics topic are averaged to provide an indication of the level of difficulty of each mathematics topic in each participating country, relative to the difficulty of other mathematics topics in the same country.

Figure 4.3 shows the average item difficulty for each mathematics topic in each country. It is important to note that the overall mean difficulty for each country has been set to 0 to allow for comparisons among nations.

The data pictured in Figure 4.3 indicate that Algebra and Measurement items are generally more difficult than Data, Geometry, and Number items in all

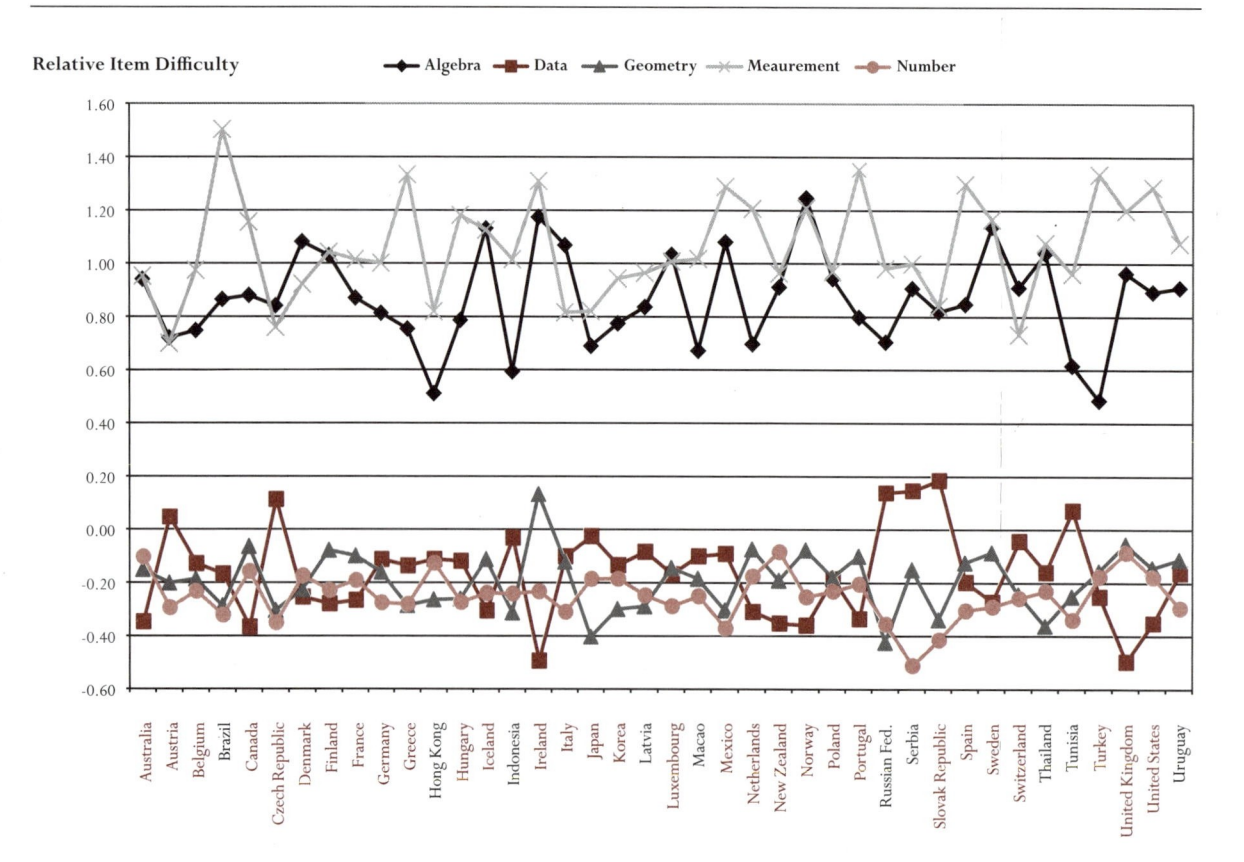

Figure 4.3 ■ **Relative difficulty by mathematics topic by country**[1]

1. Please note that mean difficulty for each country across all mathematics items is set to 0.

countries. Moreover, there is little variation across countries in average item difficulty for Number items. Greater variability is observed across countries in average item difficulty for Algebra, Measurement, and Data items. This is expected, as every country covers Number topics in earlier grades of schooling, while there are greater differences in the grade levels at which Algebra, Measurement, and Data are introduced and taught. The mean and the standard deviation of topic item difficulty across countries are given in Table 4.1.

To make grouping of the countries clearer, the relative easiness/difficulty of the topics within the countries are calculated. The topic is defined being relatively difficult (D) if the average difficulty for the country illustrated in Figure 4.3 is half a standard deviation (or more) larger than the mean across the countries provided in Table 4.1. Similarly, the topic is defined being relatively easy (E) if the average difficulty for the country illustrated in Figure 4.3 is half a standard deviation (or more) smaller than the mean across the countries provided in Table 4.1.

Algebra and Measurement are relatively more difficult, while Data, Geometry and Number are easier.

Table 4.1
Mean and standard deviation of relative topic difficulty across countries

Topic	Topic difficulty across countries Mean (SD) in logits
Number (32 items)	-0.25 (0.09)
Algebra (7 items)	0.87 (0.18)
Measurement (8 items)	1.06 (0.19)
Geometry (12 items)	-0.18 (0.12)
Data (26 items)	-0.17 (0.17)

Table 4.2 provides a summary of the four groups of countries identified at the beginning of this chapter. The full information about relative difficulty of the traditional topics can be found in Annex A3.

Across domains, one can identify different country groups.

Table 4.2 shows that for all English-speaking countries the topic Data is a relatively easy topic and for all of them, except Ireland, the topic Number is a relatively difficult topic. As in the factor analysis, Australia and New Zealand are more similar to each other than to other English-speaking countries. The United Kingdom is similar to Ireland on the one hand and to Canada on the other.

For example, data is relatively easy for English-speaking countries …

For the Northern European countries Table 4.2 also provides further insight in where exactly the similarities lie. For all of them Algebra is a relatively more difficult topic. For all except Denmark, Data is relatively easy and Geometry is relatively difficult. Noticeably, the only similarity Denmark shows with other Scandinavian countries at the topic level is the relative difficulty of Algebra.

Table 4.2
Relative easiness/difficulty of each topic within the countries

Country	Algebra	Data	Geometry	Measurement	Number
Australia		E		E	D
New Zealand		E		E	D
Canada		E	D	D	D
United Kingdom	D	E	D	D	D
Ireland	D	E	D	D	
United States		E		D	D
Finland	D	E	D		
Norway	D	E	D	D	
Sweden	D	E	D	D	
Iceland	D	E	D		
Denmark	D			E	D
Japan	E	D	E	E	D
Korea	E		E	E	D
Hong Kong-China	E		E	E	D
Macao-China	E				
Yugoslavia		D			E
Russian Federation	E	D	E		E
Latvia		D	E	E	
Czech Republic		D	E	E	E
Slovak Republic		D	E	E	E
Austria	E	D		E	E
Switzerland		D		E	

For the South-East Asia region, Algebra is a relatively easy topic and Number (except Macao-China) is a relatively difficult topic. Other traditional topics like Geometry and Measurement are relatively easy for three out of four South-East Asian countries, with Macao-China being the exception.

Some of the Central European countries and all post-communist countries share the relative difficulty of the topic Data. Further, with the exception of the Russian Federation and Yugoslavia, the topic of Measurement is comparatively easy. For the Russian Federation, Latvia, the Czech Republic and the Slovak Republic the topic of Geometry is relatively easy. Clearly, the Czech Republic and the Slovak Republic are more similar to each other than to other countries in the group and this finding was confirmed by the factor analysis discussed at the beginning of this section.

… but data is relatively difficult for Central European countries.

The association between curriculum coverage and performance on mathematics topics is further explored by examining curriculum structure at the country level. The mathematics topic Data has been chosen for a detailed discussion below, mainly because there are variations across countries in terms of the grades at which Data topics are taught, as well as the availability of curriculum information for some countries. Figure 4.3 shows that the eight countries where students found the Data topic relatively more difficult are the Slovak Republic, Serbia, the Russian Federation, the Czech Republic, Tunisia, Austria, Japan and Indonesia. In contrast, the eight countries where students found the Data topic relatively easier are the United Kingdom, Ireland, Scotland, Canada, Norway, Australia, New Zealand and the United States. For some of the countries the grades at which Data is taught can be found from the data collected in TIMSS TCMA (Mullis, Martin, Gonzalez, and Chrostowski, 2004). Table 4.3 shows the relative performance of these countries in Data (in relation to their performance on other mathematics topics) and corresponding curriculum information.

Going deeper into the curriculum structure for Data provides more insights …

Unfortunately, data are not available as part of the TCMA for Ireland, Canada, the Czech Republic and Austria. With the six remaining countries having relatively high and relatively low difficulty indices for data items, interesting patterns emerge. The data in Table 4.3 are entered by grade levels where the content associated with data are usually focused on within a country's curriculum. In some cases, especially where there is national guidance, the grade levels are entered. In other cases, where the control on focus and degree of emphasis is handled at regional or local levels, the emphasis is indicated by whether all or most students, indicated by a Y, have received this coverage by the end of Grade 8, the most able students have received it by the end of Grade 8, indicated by an M, or this has not been included for study by the end of Grade 8, indicated by an N.

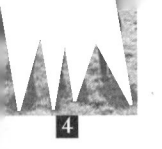

The columns in the table, labelled by Roman numerals, are associated with specific Data related outcomes:

I. Organising data using one or more characteristics using tally charts, tables or graphs

II. Sources of errors in collecting and organising data

III. Data collection methods

IV. Drawing and interpreting graphs, tables, pictographs, bar graphs, pie charts and line graphs

V. Characteristics of data sets including mean, median, range and shape of distribution

VI. Interpreting data sets

VII. Evaluating and interpreting of data with respect to correctness and completeness of interpretation

VIII. Simple probabilities including using data from experiments to estimate probabilities for favourable events

Table 4.3
Average item difficulty parameter values for Data items[1]

	Country	\multicolumn{8}{c}{Data topics (TIMSS TCMA topics)}
		I	II	III	IV	V	VI	VII	VIII
		\multicolumn{8}{c}{Grades at which the topic is taught}
Less difficult	United Kingdom[2]	K-6	8-10	6-10	6-10	5-10	6-10	8-10	4-8
	Scotland	Y[3]	M[4]	Y	Y	M	Y	M	M
	Norway	7-10	8-10	8-10	6-10	8-10	8-10	8-10	9-10
	Australia	4-7	9-10	5-8	5-8	8	9-10	9-10	7-10
	New Zealand	3-9	6-9	6-9	3-9	6-9	6-10	8-9	5-9
	United States	Y	Y	Y	Y	Y	Y	Y	Y
More difficult	Indonesia	8-9	10	11	8-11	10-12	11	11	10
	Japan	3-5	10-12	10-12	3-5	10-12	10-12	10-12	8
	Tunisia	N[5]	N	N	N	N	N	N	N
	Russian Federation	N	Y	N	Y	N	N	N	N
	Serbia	8	12	12	12	12	12	12	12
	Slovak Republic	9	9	9	7	8	9	9	7

1. From TIMSS TCMA results.
2. Note that in TIMSS, the TCMA only had data on England. In PISA, the data collected were for the United Kingdom.
3. Note that "Y" indicates that nearly all students received coverage of this topic by the end of Grade 8.
4. Note that "M" indicates that the most able students received coverage of this topic by the end of Grade 8.
5. Note that "N" indicates that the topic was not included by students by the end of Grade 8.

The first observation to be made about Table 4.3 is that the countries that find Data easier are predominantly Western countries, seven of which are English-speaking. These countries could be regarded as having similar educational traditions. In contrast, the countries that find Data more difficult are non-Western countries, three of which are from Eastern Europe where these countries also have similar educational traditions.

The second observation to be made about Table 4.3 is that the countries finding Data easier tend to teach the topic from an earlier grade. For example, in England, all of the eight Data topics are introduced at the K-8 level in schools. In New Zealand, seven Data topics are introduced in primary schools. In contrast, in Serbia, only one data topic is introduced by the intermediate level and none of the topics are introduced at this level in Tunisia.

... for example: countries where Data is relatively easier, introduce the topic from an earlier grade.

The third observation about Table 4.3 is that the countries where the Data topic is relatively more difficult appear to adopt a more linear organisational structure of the mathematics curriculum, where specific topics are only taught at specific grade levels. While there is some evidence of the same case in the countries where the Data topics were relatively less difficult, the length of the intervals of focus appear to be slightly longer. This may suggest a spiral organisational structure of curriculum, where each curriculum topic is taught across many year levels (*e.g.* 3 to 5 year levels). Consequently, in Japan, for example, students who were taught Data Topic I in Grades 3 to 5 may have forgotten about this content domain by the time they reach age 15 (Mullis, Martin, Gonzalez, and Chrostowski, 2004).

The analyses carried out in this section provide some evidence linking student performance to instruction. Countries where students have received more instruction on a mathematics topic tend to perform better in that topic relative to their performance on other topics.

PERFORMANCE AND GRADE LEVELS

If instruction has a significant impact on student performance, then one would expect some differences in performance between students from different grade levels. Clearly, one would expect students from higher grade levels to perform better, on average, than students from lower grade levels. Figure 4.4 shows the relative performance of students from different grades for a number of countries randomly selected for the purpose of illustration.

Students from higher grades perform better.

Generally speaking, for all countries that have multiple grades in the PISA study, students from higher grades performed, on average, better than students from lower grades although the magnitudes of the differences in performance between adjacent grades varied between countries. This, of course, is expected, as one additional year of schooling must increase students' performance level. Nevertheless, this finding offers further evidence that instruction is closely related to performance.

Figure 4.4 ■ **Average performance by grade for four participants**

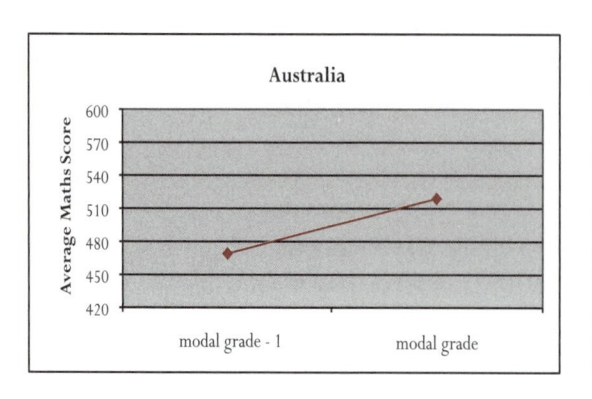

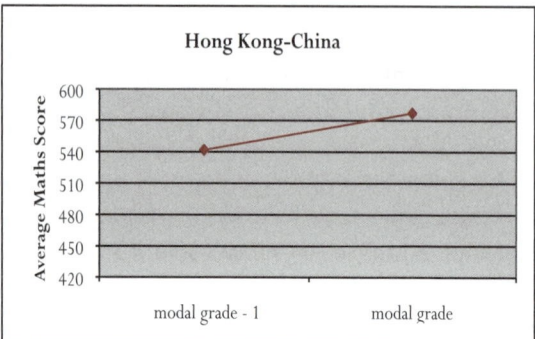

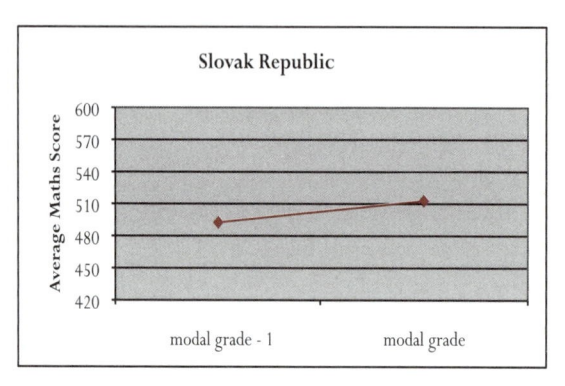

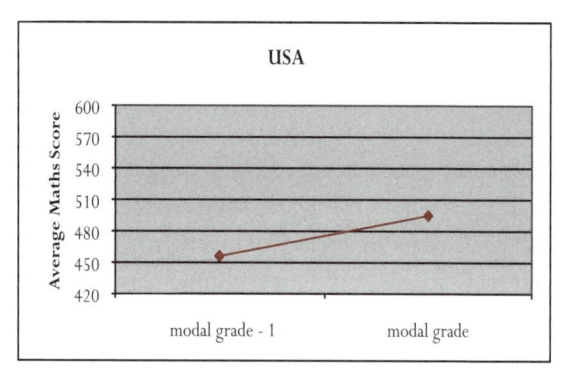

An analysis of the performance differences across grades shows that student performance is closely related with instruction.

Whether a country adopts a spiral structure of curriculum or a linear structure, there will be some variations in the topics taught at each grade level. Consequently, some items may show Differential Item Functioning (DIF) for students because of the inclusion of students from different grade levels. That is, for some items, students with the same ability will be likely to have different probabilities of success if they are from different grade levels. To test this hypothesis, DIF analyses were carried out for a selected number of countries where there were substantial numbers of students from different grades. The items for which lower grade students were most disadvantaged were identified. These items were further examined in terms of content and, where possible, in terms of national curriculum. Table 4.4 shows the results.

A number of observations can be made from the analyses of Grade DIF. First, the magnitudes of DIF across grades within countries are generally less than the magnitudes of DIF across countries. For a particular item, the maximum difference between item difficulty parameters across countries is typically between 1 and 2 logits (see variations of item difficulties in Figure 4.1, for example). In comparison, within a country, the maximum difference between item difficulty parameters between two grades is around 0.5 logit.

Second, of the eighteen entries in Table 4.4, eight are Algebra, although only four of these eight entries are different items. In particular, the Algebra item THE

THERMOMETER CRICKET Q2 [M446Q02] appears in four of the six countries, as showing differential item functioning between two adjacent grades. This item is also the most difficult item among all mathematics items in PISA 2003. As Algebra is generally not taught until later years of schooling, it is not surprising that there is a greater chance that these items show DIF across grade levels. That is, when there are two students with the same overall mathematics ability, the student from a higher grade will have a higher probability of success on an algebra item than a student from a lower grade. From this point of view, the

Table 4.4
Items identified with grade DIF for countries with multiple grades

	Three items where lower grade is disadvantaged most	Mathematics topic	Subtopic	Grades taught	Item difficulty calibrated for the country (logits)	Item difficulty difference between two grades (logits)
Australia	THE THERMOMETER CRICKET Q2	Algebra	Equations	9-10	3.22	0.46
	WALKING Q01	Algebra	Equations	9-10	1.26	0.33
	EXCHANGE RATE Q02	Number	Ratios	9-10	-1.11	0.32
Austria	RUNNING TRACKS Q03	Measurement	Formulas	N[1]	1.63	0.32
	THE THERMOMETER CRICKET Q2	Algebra	Equations	N	3.06	0.29
	WALKING Q01	Algebra	Equations	N	0.57	0.27
Hong Kong-China	GROWING UP Q3	Data	Interpretation	10-11	1.15	0.74
	THE BEST CAR Q01	Algebra	Equations	7-9	-1.92	0.34
	THE BEST CAR Q02	Algebra	Equations	7-9	1.17	0.33
Russian Federation	HEIGHT Q01	Data	Interpretation	N	-0.06	0.36
	THE THERMOMETER CRICKET Q2	Algebra	Equations	N	3.1	0.36
	CUBES Q01	Data	Represent	N	-0.55	0.31
Slovak Republic	CHOICES Q01	Number	Patterns	9	-0.12	0.32
	HEIGHT Q02	Data	Statistics	8	2.13	0.28
	THE FENCE Q01	Measurement	Formulas	9	1.34	0.28
United States	THE THERMOMETER CRICKET Q2	Algebra	Equations	N	3.26	0.5
	CUBES Q01	Data	Represent	N	-0.99	0.36
	CARBON DIOXIDE Q01	Data	Represent	N	0.62	0.36

1. Note that "N" indicates that no information is available.

students from lower grades are disadvantaged, owing to the fact that they have not received as many instructional lessons on Algebra as students from higher grades. This is often referred to as the OTL, or opportunity to learn, factor.

Third, in addition to Algebra items, some Data items also exhibit DIF across grade levels. In particular, Hong Kong-China, the Russian Federation, the Slovak Republic and the United States all have Data items showing grade DIF. In Table 4.3 students in the Russian Federation and the Slovak Republic found Data items more difficult than students in most other countries. This is an indication that Data is generally not taught until higher grades in these countries. Consequently, it is not surprising that some Data items also show grade DIF.

Fourth, for countries where there is information about curriculum structure, it appears that grade DIF items relate to topics that are taught at higher grade levels. For example, a Number item is identified as exhibiting grade DIF for Australia. This Number topic is only taught at Grade 9. Similarly, for Hong Kong-China, the Data item GROWING UP Q3 is found to exhibit grade DIF. The content area for this item is taught at Grades 10-11.

Finally, grade DIF items tend to be more difficult items. The average item difficulty for the items in Table 4.3 is 0.99 logit, where the average item difficulty for the whole set of mathematics items is 0 logit for each country.

In summary, the identification of grade DIF items provides support for the hypothesis that student performance is closely linked to instruction. Moreover, the identification of specific grade DIF items for each country can shed some light on the curriculum structure in the country and provide the basis for possible intervention strategies if necessary. However it must be noted that PISA is designed primarily as an age-based survey, so the presence of multiple grades within a country is not controlled. As such, the study design of PISA does not lend itself to in-depth analysis of grade differences for all countries.

COMPETENCY CLUSTERS AND MATHEMATICS PERFORMANCE

In PISA mathematical competencies are organised into three clusters (see Chapter 2). The PISA 2003 mathematics assessment included 26 questions in the *reproduction* competency cluster, 40 questions in the *connections* competency cluster, and 19 questions in the *reflection* competency cluster.

Across competency clusters, Reproduction was the easiest and Reflection the most difficult.

The relative difficulties of questions included within each competency cluster are presented by country in Figure 4.5. These statistics show that questions in the *reproduction* competency cluster were on average the easiest and those in the *reflection* competency cluster were on average the most difficult. This relationship for difficulty of questions within competency clusters holds for all of the participating countries.

Table 4.5
Mean and standard deviation of question difficulty by competency cluster across countries

Competency cluster	Number of questions included	Difficulty of questions included across countries (in logits) Mean (SD)
Reproduction	26	-1.00 (0.06)
Connections	40	0.26 (0.03)
Reflection	19	0.82 (0.09)

Moreover, there is little variation across countries in average question difficulty for all competency clusters. The mean and the standard deviation of question difficulty for each competency cluster across countries are given in Table 4.5.

It was shown earlier in this chapter (see Table 4.3) that Algebra and Measurement questions are significantly more difficult than Number, Geometry and Data across all countries. Table 4.6 shows the distribution of PISA questions by the traditional mathematics topic and by competency cluster. The competency clusters include questions from each of the traditional mathematics topics, although the *reproduction* competency cluster does not include Measurement questions.

Figure 4.5 ■ **Average question difficulty by competency cluster in participating countries**

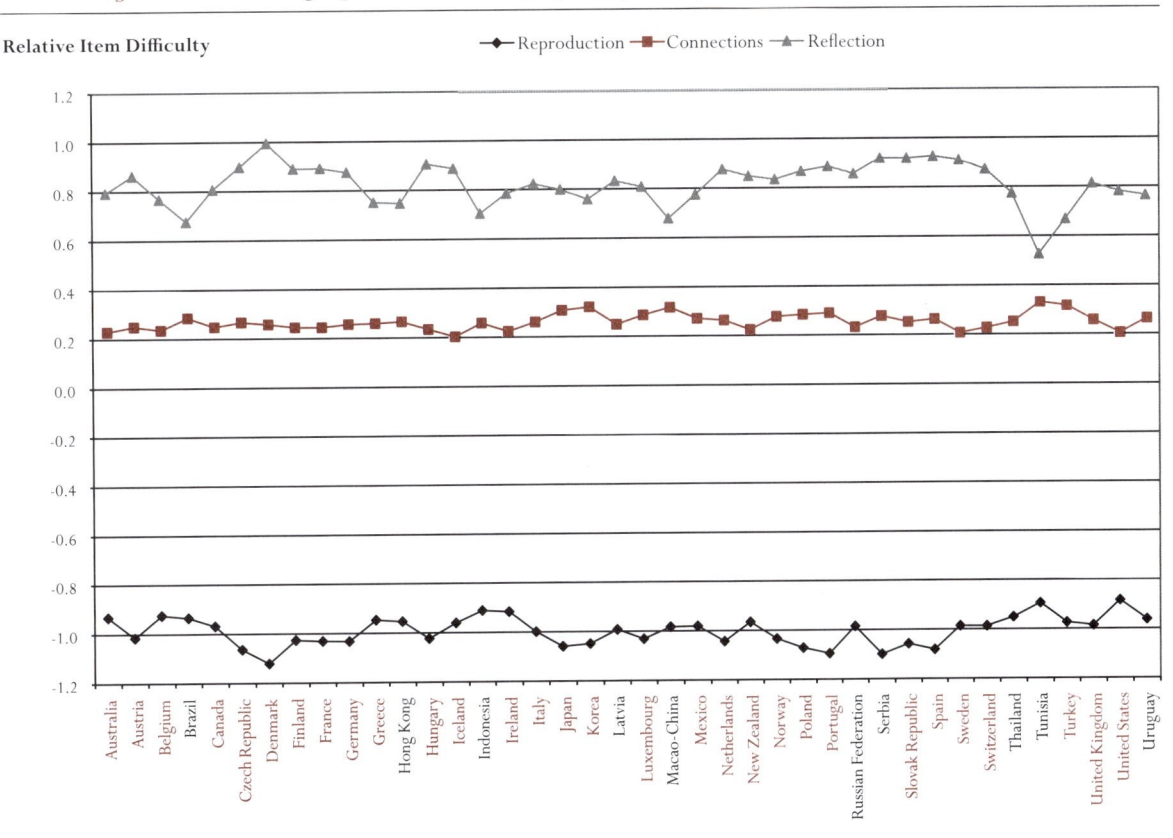

Table 4.6
Questions in competency clusters by traditional mathematics topic

	Algebra	Data	Geometry	Measurement	Number
	\multicolumn{5}{c}{Percentage of questions (number of questions)}
Reproduction	28.6% (2)	34.6% (9)	25.0% (3)	0.0% (0)	37.5% (12)
Connections	28.6% (2)	42.3% (11)	58.3% (7)	75.0% (6)	43.8% (14)
Reflection	42.9% (3)	23.1% (6)	16.7% (2)	25.0% (2)	18.7% (6)
Total	100% (7)	100% (26)	100% (12)	100% (8)	100% (32)

Both competency clusters and traditional domains are related to the difficulty of PISA questions.

Figure 4.6 suggests that the difficulty of the content in PISA questions is attributed to the traditional mathematics topic, as well as to the competency clusters. That is Algebra questions are more difficult on average within each competency cluster. The same applies to Measurement. On the other hand, the *reflection* competency cluster is more difficult within each traditional topic except Geometry. Interaction between competency clusters and traditional topics is most likely due to this Geometry effect and the absence of Measurement questions in the *reproduction* cluster.

Algebra questions are more likely to involve such competencies as symbols and formalism (see Chapter 2) which is a defining competency of *mathematical literacy*. It relates to the ability to handle and work with statements containing symbols and formulas, as for example, in THE BEST CAR – Question 1. This is the easier of two Algebra questions in the *reproduction* competency cluster and

Figure 4.6 ■ **Average question difficulty by competency cluster and by traditional topic**

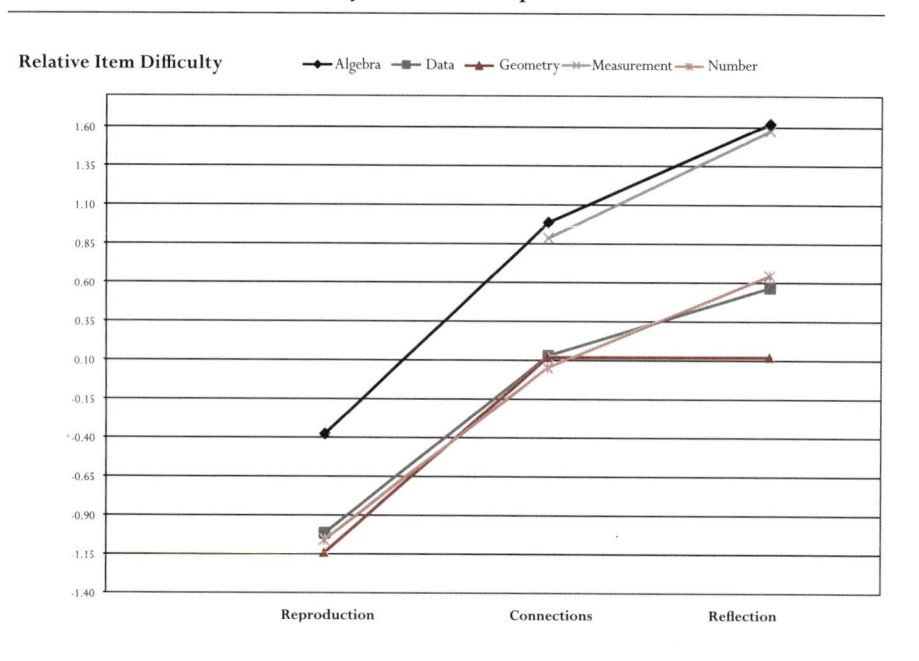

requires number substitution into a given formula. The second of these questions is WALKING – Question 1 which is much more difficult as it requires both substitution into a given equation and solving the equation (see more about Algebra questions in Chapter 5).

CONTEXT AND MATHEMATICS PERFORMANCE

PISA's focus on *mathematical literacy* reflects an increasing concern about how well students can apply mathematics to solve real-life problems. Therefore, PISA mathematics questions are contextualised, reflecting different aspects of the real world such as travel, sport, media, modern communication and science, but also intra-mathematical contexts that reflect part of students' experience of mathematics in school.

PISA questions presented in a personal context were easier than those in other contexts.

PISA questions are classified into four different contexts or situations: *educational and occupational*, *scientific*, *personal* and *public* (see Chapter 2). Figure 4.7 presents the average question difficulty in each of the four contexts for each country. As in the previous sections, the mean difficulty for all mathematics items is set to 0 for all countries.

Overall, the easiest for all countries were questions presented in a *personal* context, and for the majority of countries the most difficult were questions presented in a *scientific* context. The standard deviations (see Table 4.7) are

Figure 4.7 ■ **Average question difficulty by context in participating countries**

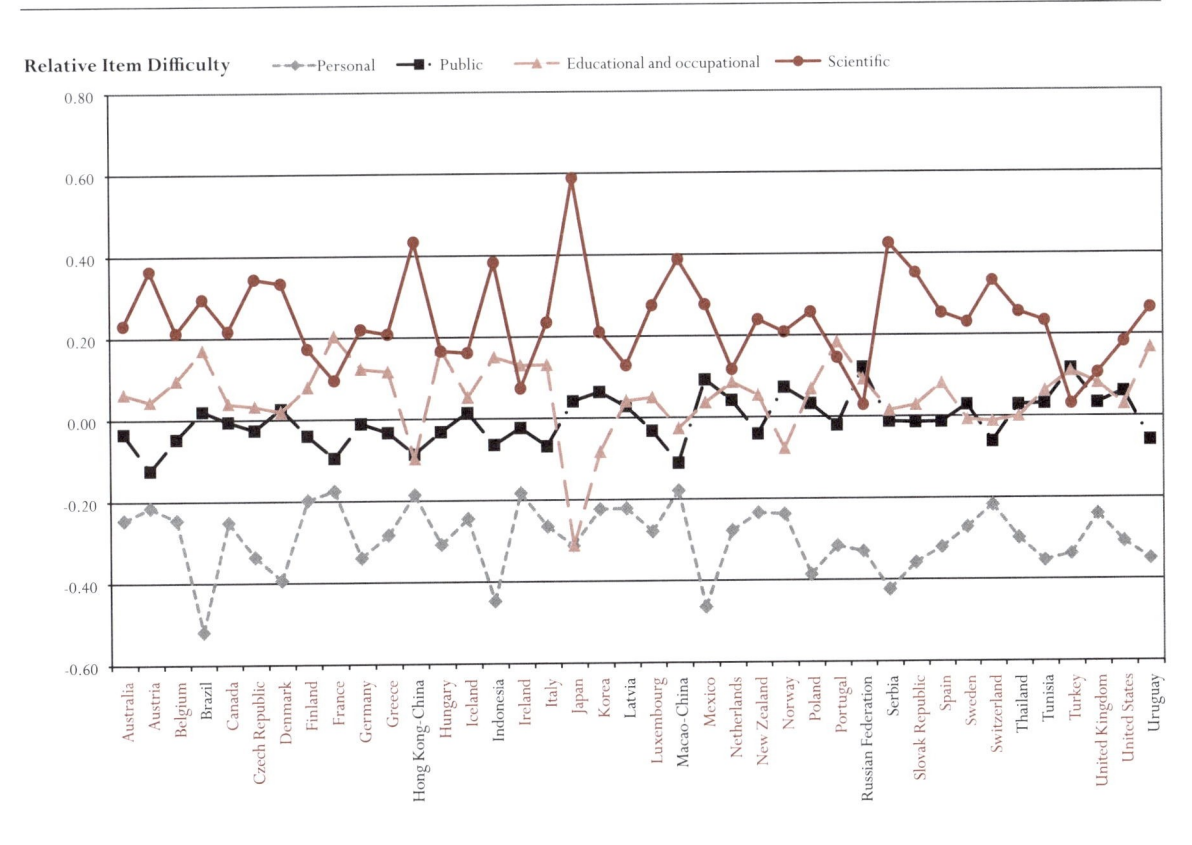

Table 4.7
Mean and standard deviation of question difficulty by context across countries

Context	Number of questions included	Difficulty of questions included across countries (in logits) Mean (SD)
Scientific	18	0.24 (0.11)
Educational and occupational	20	0.05 (0.09)
Public	29	0.00 (0.06)
Personal	18	-0.29 (0.08)

slightly higher than those observed for competency clusters but still relatively small.

Multiple comparisons confirmed that overall across 40 countries questions which were presented in a *personal* context were easier than all other questions. Also, questions presented in a *scientific* context were more difficult than all other questions except those presented in an *educational and occupational* context. The differences are small, but statistically significant. There is no difference in difficulty between questions presented in a *public* context and questions presented in an *educational and occupational* context (see Table 4.8).

Table 4.8
Multiple comparisons of question difficulty by context across countries (using Bonferroni adjustment)

(I) Context	(J) Context	Mean Difference (I-J)	Std. Error	Sig.	99% Confidence Interval Lower Bound	99% Confidence Interval Upper Bound
Personal	Public	-0.29	0.06	0.00	-0.48	-0.10
	Educational and occupational	-0.35	0.06	0.00	-0.55	-0.15
	Scientific	-0.54	0.07	0.00	-0.74	-0.33
Public	Educational and occupational	-0.06	0.06	1.00	-0.24	0.12
	Scientific	-0.25	0.06	0.00	-0.43	-0.06
Educational and occupational	Scientific	-0.19	0.06	0.02	-0.39	0.02

Differences in question difficulty by context are not significantly different for most countries, although, for low achieving countries questions presented in a *personal* context were relatively easier (Mexico and the partner countries Brazil and Indonesia) and for Japan questions presented in a *scientific* context were relatively more difficult.

CONCLUSION

This chapter examined the performance of countries in terms of their relative strengths and weaknesses in different traditional curriculum topics. It was found that the observed differences across countries in their performance patterns could be linked to curriculum and instruction. In particular, English-speaking countries have similar performance patterns. Students in English-speaking countries tend to perform relatively better on Data questions. Where available, evidence from TIMSS shows that instruction about Data is introduced in the early grades of schooling in these countries. In contrast, the Czech Republic, Japan, the Slovak Republic and the partner countries/economies Hong Kong-China, Serbia and the Russian Federation often appear in the same groups as better performing countries in Algebra, Geometry, Measurement and Number.

Performance in PISA is related to curriculum and instruction.

While these findings are similar to those of some earlier studies (*e.g.* Zabulionis, 2001), there are some differences. In particular, Eastern European countries such as the Czech Republic, Hungary, Poland, the Slovak Republic, and the Latvian-speaking part of the partner country Latvia seem to be moving closer in their performance patterns to those of Western European countries than they were several years ago. This could reflect a gradual change in direction in curriculum structure in these countries.

Patterns of performance emerge across countries …

An examination of the performance patterns across grades within a country shows that there are some differences across grades, particularly for topics taught only in higher grades. However, the differences in performance patterns across grades are small as compared to performance pattern differences across countries. Nevertheless, the link between instruction and student performance is again evident.

… grades …

As expected, questions from the *reproduction* competency cluster on average are easier than the questions from the *connections* competency cluster, while questions in the *reflection* competency cluster are the most difficult of the three. This is true for all countries with little variation.

… competency clusters …

Regarding the context, *personal* questions are on average the easiest and *scientific* questions are on average the most difficult, and this is true for all countries, although the differences are small. The challenge for educational practitioners is to make *scientific* questions more attractive for students.

… and contexts.

In conclusion, the analyses carried out in this chapter show that PISA results can provide useful information about student performance and instruction.

Further, curriculum structures have a significant impact on student performance. The results from this chapter provide a starting point for an examination of curriculum structures in each country, as well as an assessment of the relative merits of different curriculum designs. For example, if it is deemed important that future citizens should have a sound knowledge of statistical methods for the dissemination of information and data, then the Data topic should be introduced earlier and emphasised more in the curriculum. On the other hand, if there is a need for better preparation for tertiary science, economics, statistical, and technical studies and better understanding of future citizens of various dynamic of processes, then Algebra and the study of functions should be emphasised more. Consequently, the results from this chapter can provide a basis for a re-evaluation of curriculum designs in each country.

As PISA carries out data collection every three years, the analyses carried out in this chapter can be repeated, so that trends in the performance patterns of countries by curriculum topics can be monitored and cross-checked with curriculum changes in each country.

Note

1. For a general description of cluster analysis see, for example, Anderberg (1973).

The Roles of Language and Item Formats

This chapter focuses on factors other than the three Cs (mathematical content, competencies and context) that influence students' performances. Just as countries differ, students' experiences differ by their individual capabilities, the instructional practices they have experienced, and their everyday lives. The chapter examines some of these differences in the patterns of performance by focusing on three factors accessible through data from PISA 2003: language structure within PISA 2003 assessment items, item format, and student omission rates related to items.

INTRODUCTION

What role does the wording of the problems themselves play in the PISA findings for mathematics? There is a vast literature detailing the importance of language factors in mathematics learning (Ellerton and Clements, 1991). The literature on performance assessments suggests that the use of language in test questions can influence the difficulty of the question and therefore students' performance on assessments (O'Leary, 2001; Routitsky and Turner, 2003). In this chapter, different aspects of the use of language are investigated, including the length of the text (number of words) and therefore the amount of reading required to understand the question.

A question's language and format influences whether students answer correctly or …

Question format also has a potential to influence students' performance through its structure and response demands (O'Leary, 2001; Routitsky and Turner, 2003). The types of questions asked and the types of responses required by students vary considerably with the PISA mathematics assessment. Some questions require students to provide one simple answer, such as just a number, or select an answer from a range of possible responses. Other questions require students to provide an answer and explain why or justify how they came to their particular conclusion. The reasoning demands and response constraints that each question type can have on student performance varies across countries due to their differences in curriculum, instructional practices and students' everyday experiences. The analysis in the second part of this chapter will investigate the relationship between the different types of questions used in PISA mathematics and their difficulty.

… whether they even attempt to answer it.

Another related issue in international assessments is the difference in omission rates, meaning the percentages of students who do not attempt to answer questions. Omission rates consider patterns of non-response that occur even after student data is conditioned for non-completion, due to time constraints of the testing situation. Beyond time, the responsiveness of students (patterns of missing values) can depend on item characteristics that can be intentionally varied or controlled such as item difficulty, item format, the mathematical content involved, the context of the item, the level of reading demand involved and the amount of information in the stimulus (Jakwerth, Stancavage and Reed, 1999). Of course, omission rates may also be influenced by factors other than specific item characteristics that are outside the control of a teacher or assessment designer, for example cultural factors; however such factors lie outside the scope of this report. The chapter concludes with an analysis of patterns of differences in student omission rates on PISA mathematics assessments.

THE USE OF LANGUAGE IN PISA MATHEMATICS QUESTIONS AND STUDENT PERFORMANCE

As PISA is the first large-scale international study to assess *reading literacy*, *mathematical literacy* and *scientific literacy*, it is particularly important to also consider the use of language in contextualising the questions. As in Chapter 4, the focus in this chapter is on difficulty of the groups of questions within each country. Here, the questions are grouped by number of words.

For the purpose of this analysis, all the PISA 2003 mathematics questions were closely analysed and classified according to the number of words used in the question. This was done using the English language version of the test questions by counting all of the words used in both the stimuli and the questions. However, the calculation of the number of words for each question was not straightforward in all cases. Questions where this process was straightforward include the so-called "single-question units" where there is no clear distinction between the stimulus and the question, the question was presented as a whole (*e.g.* CUBES Q1 and STAIRCASE Q1). Further, for many of the questions belonging to a unit with more than one question included, it was necessary for students to read the information in the stimulus in order to answer the question (*e.g.* both questions in the unit WALKING or two out of three questions in the unit GROWING UP Q2 and Q3). However, there were a few questions where information within the question itself was sufficient for students to answer the question without reading the stimulus, such as GROWING UP Q1.

The number of words in a question measures its reading load.

It can be expected that different students would use different reading strategies. Hopefully, all students would read the stimulus before attempting to answer the questions. For GROWING UP Q1, reflective students may have laboured over the stimulus unnecessarily. However, careful reading of the stimulus would save them time when answering GROWING UP Q2 and Q3. Another strategy would be to quickly look through the stimulus, answer the first question, and then return to the stimulus again when answering the second and third questions. These strategies would influence the time required for each question, but not necessarily the difficulty of the questions. Whether the student would read the stimulus of GROWING UP carefully or just looked through it quickly, GROWING UP Q1 would still require only a simple subtraction of two numbers given in the question itself.

After careful consultation and consideration, it was decided that if information in the stimulus was required for students to answer the question, the number of words is counted as the sum of the number of words in the stimulus and the number of words in the question itself. If the information within the question is sufficient for students to be able to give the answer, then the number of words is counted as only the number of words in the question, including words in any graphic elements and words used to formulate answers for multiple-choice questions, if applicable.

WORD-COUNT AND QUESTION DIFFICULTY ACROSS COUNTRIES

Using the methodology detailed above, the correlation between the number of words and the question difficulty in OECD countries was 0.28. To find out more about the relationship between the number of words and the difficulty of the items, all items were divided into three categories: short (50 words or less), medium (between 51 and 100 words), and long (more than 100 words). Of course, a text with more than 100 words in the English version will have more or fewer words according to the language into which it is translated. The English version is used as a basis of text length to approximate the measure of the real length of the texts presented to students. The three categories represent a hierarchy of the texts according to their lengths (word count).

Short and medium questions are equally difficult and long questions are the most difficult.

Figure 5.1 shows the average relative difficulty of questions in each of the three word-count categories for each country. The short and the medium-length questions are on average of similar difficulty, while the long questions are significantly more difficult across all countries. Variation between countries within the medium category is small, while variation is slightly higher for the long and short categories (see Table 5.1)

So, longer questions are on average more difficult in all countries, but does the different difficulty of longer and short/medium-length questions explain performance differences across countries? This was investigated, and no significant differences were found for the majority of the countries (See Annex A4).

The difficulty of questions for each country (in logits centered at 0) was defined as the dependent variable and word-count and country were defined as independent

Figure 5.1 ■ **Average relative difficulty of questions within each word-count category for each country**

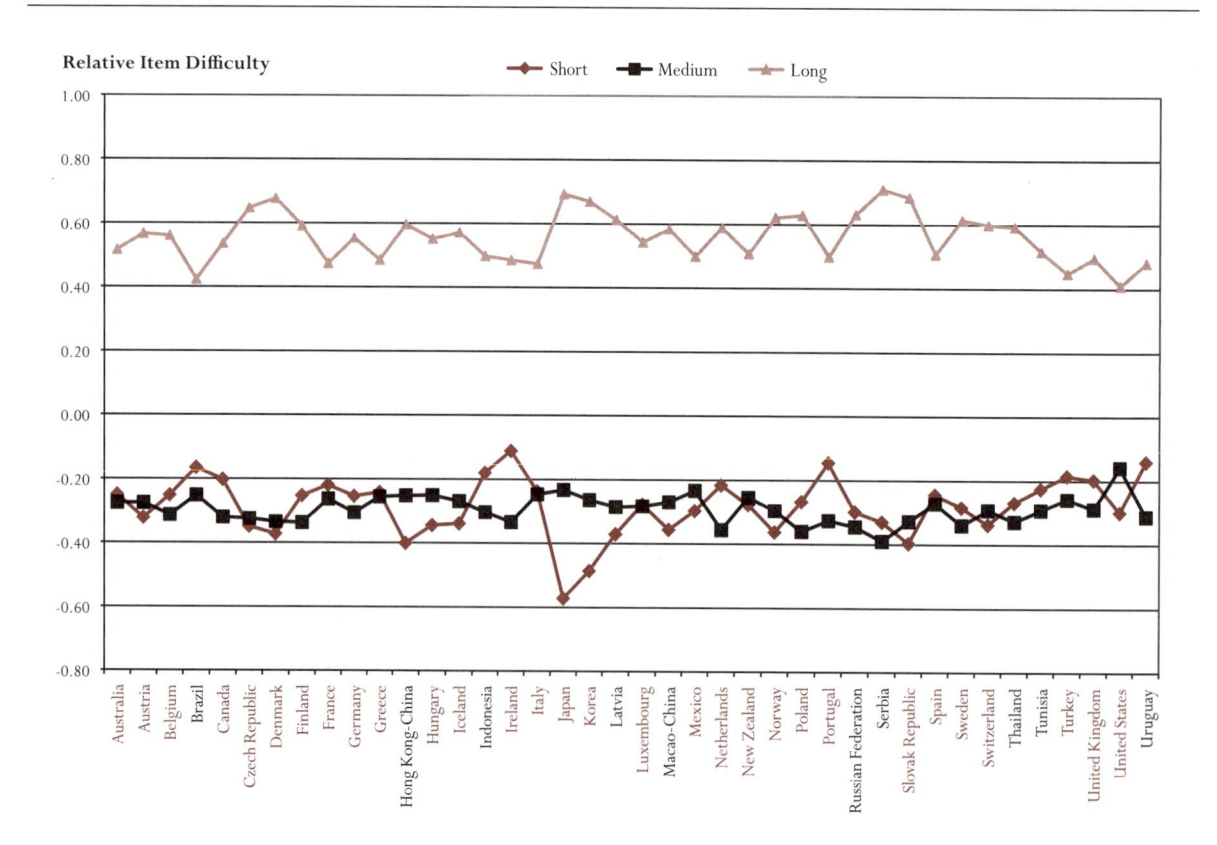

Table 5.1
Mean and standard deviation of difficulty of questions in each word-count category across countries

Word-count group	Number of questions included	Difficulty of questions included across countries (in logits) Mean	(SD)
Short	21	-0.28	(0.09)
Medium	35	-0.29	(0.04)
Long	29	0.56	(0.08)

variables or factors (for a description of ANOVA see, for example, Rutherford, 2001). The results of this analysis (see Annex A4) show that while word-count categories across countries are significantly different, the interaction between countries and word-count is not significant. That is, between countries the variation within each category of word-count is indeed small. So there is really very little variation within each of the three word-count groups between countries (see Annex A4 for details). However, there were a few exceptions. The short questions are relatively easier for students in Korea and Japan (mean difficulty -0.49 and -0.57 logits respectively; both means are more than two standard deviations away from the overall average for the short questions). For the partner country Serbia the long questions are relatively more difficult (0.71 logits) while the medium-length questions are relatively easier (-0.39 logits). Finally, for the United States the long questions are relatively easier (0.41 logits) while the medium-length questions are relatively more difficult (-0.16 logits).

These few exceptions cannot be easily attributed to particular instructional and cultural differences. However, it is possible that this apparent effect of question length is confounded by other question characteristics that have already been analysed in Chapter 4, such as the mathematical content, the context in which the question is presented and the mathematical competencies required to answer the question. These factors are examined in the next section. It was also found that long questions are on average more difficult than medium-length (and short) questions, while there is no significant difference between the mean difficulty of medium-length and short questions (see Annex A4, Table A4.5).

WORD-COUNT AND THE CONTEXT IN WHICH A QUESTION IS PRESENTED

To what extent is the amount of reading involved in the question connected to the context in which that question is presented? This was investigated for all countries overall and results show that there is a small interaction between the context and the number of words used in the question, and that of these two factors, it is the number of words used that contributes more to the difficulty of the question. However, there are no differences among countries in this respect. The check for interaction between word-count and context in relation to question difficulty was also investigated through a full factorial analysis of variance where the dependent variable was defined as question difficulty for each country (in logits centered at 0) and a factor for context was added (See Annex A4, Tables A4.3 and A4.4 for full results).

A question's context and length are related.

The results of this analysis show that there is a small but significant interaction between context and word-count which accounts for about 4% of the overall variance. At the same time, word-count as a main effect accounts for about 12 % of the overall variance and context as a main effect accounts for about another 4% of the overall variance. This can be interpreted to mean that word-count is a more important predictor of item difficulty than context. There were no significant interactions between context and country or between word-count and country, which means that country differences within each of the categories are insignificant.

Figure 5.2 illustrates the interaction between word-count and context. It shows quite different behaviour in the *educational and occupational* context area compared to all other areas in relation to the item difficulty for different categories of word-count.

The distribution of word-count for each context area was calculated to find out why the *educational and occupational* context area behaves differently compared to all other areas in relation to the item difficulty for different categories of

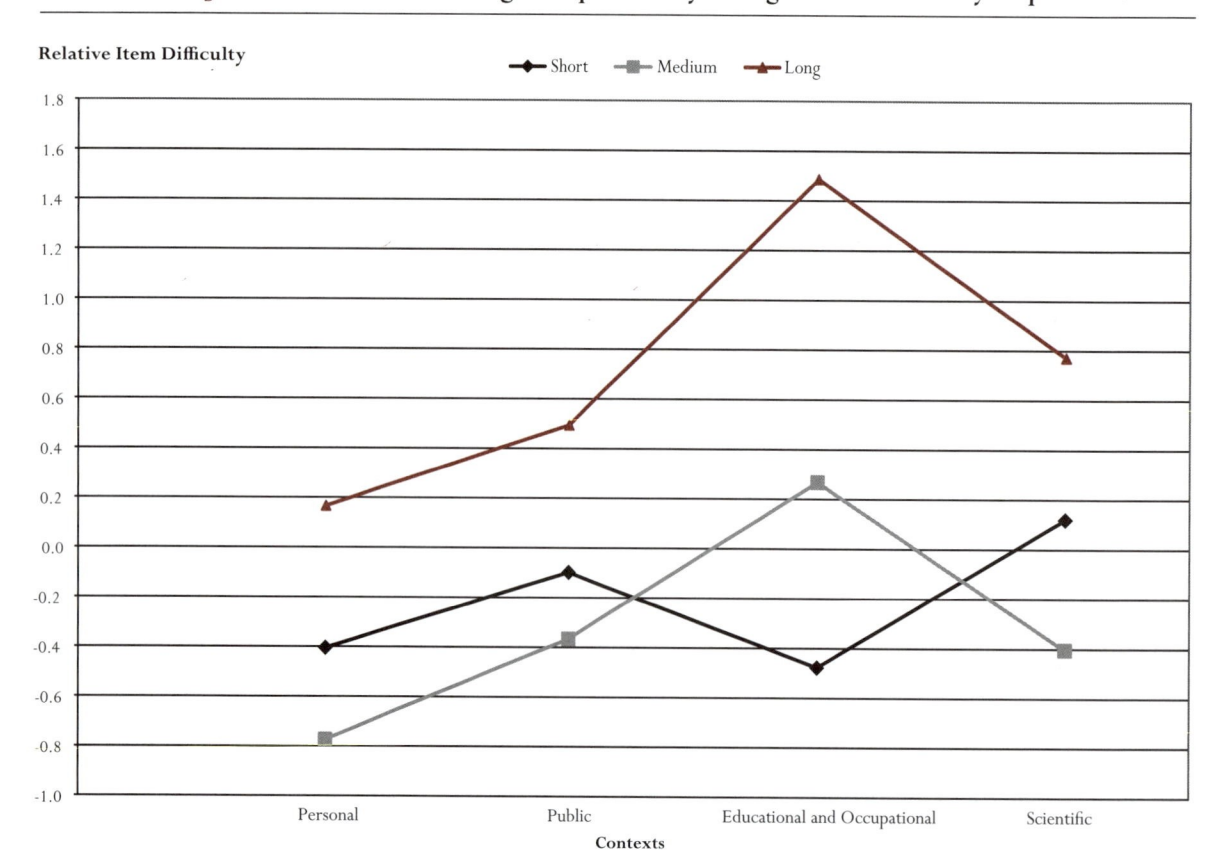

Figure 5.2 ■ **Context and length of question by average relative difficulty of questions**

Table 5.2
Item distribution by context by word-count

Item group by word-count	Context			
	Educational and occupational	Personal	Public	Scientific
	Percentage of items in each category of word-count (number of items)			
Short	45% (9)	17% (3)	24% (7)	11% (2)
Medium	45% (9)	39% (7)	41% (12)	39% (7)
Long	10% (2)	44% (8)	34% (10)	50% (9)
Total	100% (20)	100% (18)	100% (29)	100% (18)

word-count. Table 5.2 shows that the distribution of items in the *educational and occupational* context area is quite different from the distribution of items in the other context areas. It has only two long items (10% of total items in the context area); while other content areas have eight to ten long items (34-50%).

It is worth noting that it is the number of items in each of the word-count categories, and not the average number of words, that explains the interaction between word-count and context. Table 5.3 shows that the average number of words in each of the word-count categories in the *educational and occupational* context area is quite similar to the average number of words in each of the word-count categories for the other context areas.

Therefore it is more likely that it is the distribution of items by word-count, and not the average number of words, that is responsible for the interaction between word-count and context.

Table 5.3
Average number of words by context by word-count

Item group by word-count	Context			
	Educational and occupational	Personal	Public	Scientific
	Average number of words in each category			
Short	32	34	30	30
Medium	74	74	65	86
Long	138	144	115	143

WORD-COUNT AND COMPETENCIES REQUIRED TO ANSWER THE QUESTION

The methodology used in this section is the same as in the previous section. First, full factorial analysis of variance was performed. As in the previous section, item difficulty for each country (in logits centered at 0) was defined as the dependent variable. Additional factors were country, word-count, and competencies. The results of this analysis (see Annex A4, Table A4.2) show that there is a small but significant interaction between competencies and word-count that accounts for about 2% of the overall variance. At the same time, competencies as a main effect account for about 23% of the overall variance, and word-count as a main effect accounts for another 5%. There were no significant interactions between competencies and country or word-count and country, which means that country differences within each of the categories are insignificant. Unlike in the previous section, it is not the word-count that is responsible for most of the variance, but competencies.

A question's competency cluster adds more to its difficulty than its word-count.

Figure 5.3 shows the relationship described above. In particular it shows that the differences between the *reproduction* cluster on the one hand and the *connection and reflections* clusters on the other hand are larger than the differences between the word-count categories within each of the competency clusters. This demonstrates the greater importance of the competencies compared to the word-count. Figure 5.3 also illustrates that the *connections* cluster behaves differently in relation to the item difficulties in each of the word-count categories. This illustrates the interaction between competencies and word-count.

Figure 5.3 ▪ **Competency clusters and length of question by average relative difficulty of questions**

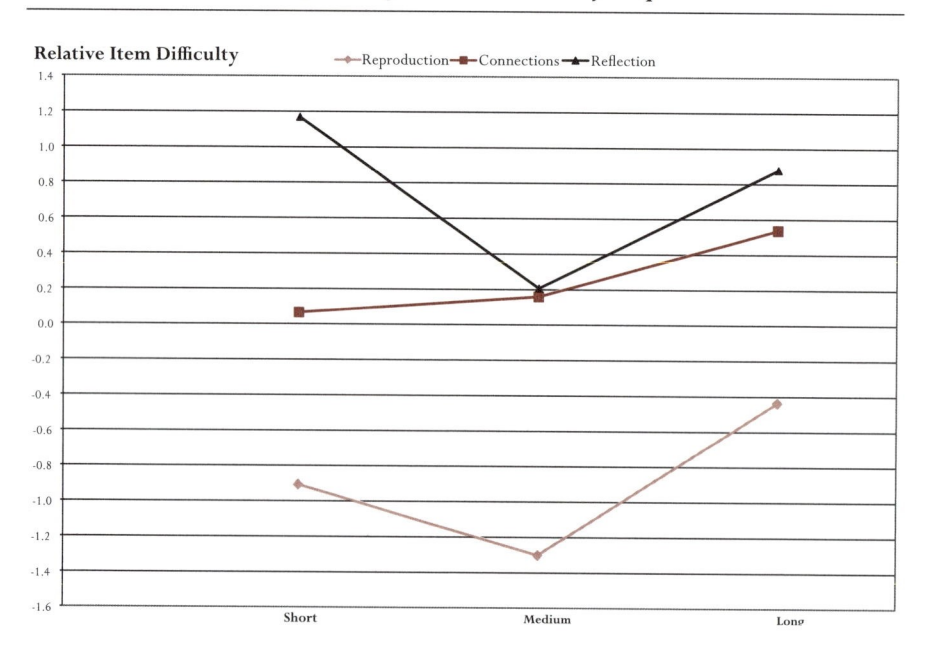

Figures 5.2 and 5.3 show that medium and short items appear to differ in average difficulty when subdivided between competency clusters or context areas. Unexpectedly, the short items look more difficult on average than the medium items. This happens in three out of four context areas (Figure 5.2) and in two out of three competency clusters (Figure 5.3). This gives us some indication that the shortest items are not always the easiest. It is possible that sometimes more explanations in the stimulus (provided that the explanations are not too long) make items easier.

Table 5.4 shows a somewhat expected item distribution within the word-count categories in each of the competency clusters. In the *reproduction* cluster only 16% of the items belong to the long category, while in the *connections* cluster long items make up 30% and in the *reflection* cluster they make up 68%.

Table 5.4
Item distribution by competencies by word-count

Item group by word-count	Reproduction	Connections	Reflection
	Percentage of items in each category of word-count (number of items)		
Short	42% (11)	18% (7)	16% (3)
Medium	42% (11)	53% (21)	16% (3)
Long	16% (4)	30% (12)	68% (13)
Total	100% (26)	100% (40)	100% (19)

WORD-COUNT AND CONTENT

Finally, the same methodology was applied to the content. The traditional topics described in detail in Chapter 4 were chosen as content categories rather than overarching ideas in order to make a more direct connection to traditional curriculum. The results of the full factorial analysis of variance (see Annex A4, Table A4.3) show that there is a larger interaction between content and word-count than between competencies and word-count or context and word-count. The interaction between content and word-count accounts for 9% of the overall variance. This means that in relation to word-count, the topics differ much more than competency clusters or context areas.

The interaction between content and word count is strong …

At the same time, word-count as a main effect accounts for only 3% of the overall variance while content as a main effect accounts for about 16% of the overall variance. This means that the traditional topics are more important predictors of item difficulty than the word-count, yet not as important as competencies.

… but content is a better predictor of difficulty than word-count.

As in previous sections, there were no significant interactions between context and country or between word-count and country.

Question length and difficulties vary considerably across content areas.

Figure 5.4 shows quite different behaviour of the five topics in relation to word-count. This corresponds to the high interaction between the word-count and the topics.

One observation from Figure 5.4 is that the variation of average item difficulty by word-count is higher for Data, Number and Measurement than for Algebra and Geometry. Although the number of items in each category (see Table 5.5) is not large enough to draw a conclusion, this evidence suggests that for the difficulty of the Algebra items the word-count is less important than formula manipulations and other algebraic cognitive demands. Similarly for Geometry items we can suppose that spatial cognitive demands influence item difficulty more strongly than reading demands, thus reducing the influence of the word-count variable. The results should be treated with caution given the small number of items in some combinations of word-count category and topics (see Table 5.5).

Short Measurement questions are the most difficult ones.

Table 5.5 shows a quite unequal distribution of items by word-count in each topic. On the one hand this distribution partially explains the high interaction between the content and the word-count. On the other hand it represents the relationship between the topics and the PISA framework. Nearly all Geometry items in PISA are related to two- and three- dimensional shapes, location and spatial relation, and symmetry and transformation. Thus it is not surprising that more than half of them are short items. Measurement, not surprisingly, also has only one long item.

Algebra questions are either long or medium length.

At the same time, Algebra items, when situated in a realistic context, require a somewhat wordy explanation of this situation and as a result the Algebra items do not have short items at all.

Figure 5.4 ■ **Traditional mathematics topics and length of question by average relative difficulty of questions**

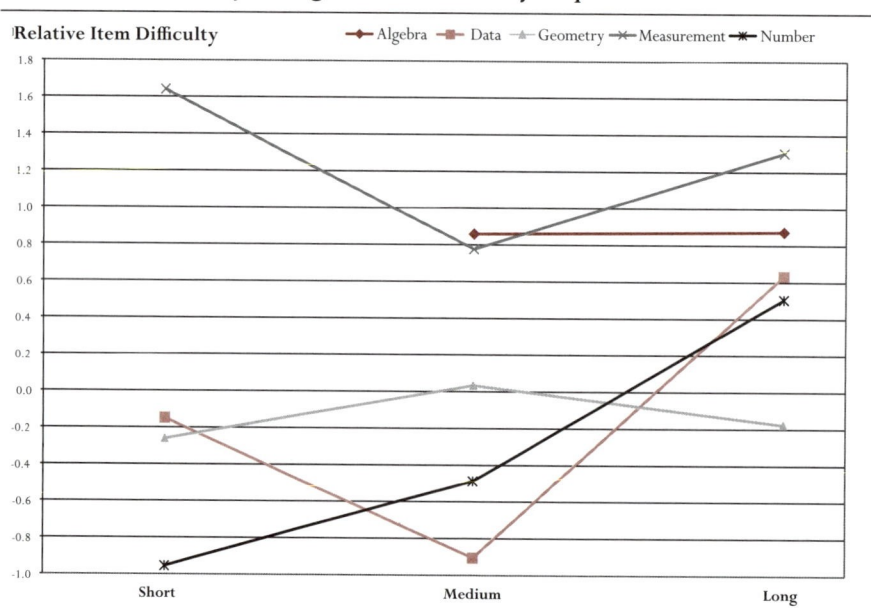

Table 5.5
Distribution of questions by traditional topic and length of question

| Word-count category | Content (traditional topic most predominantly tested) |||||
	Algebra	Data	Geometry	Measurement	Number
	Percentage of questions in each word-count group (number of questions)				
Short	0% (0)	19% (5)	58% (7)	25% (2)	22% (7)
Medium	43% (3)	42% (11)	17% (2)	62% (5)	44% (14)
Long	57% (4)	39% (10)	25% (3)	13% (1)	34% (11)
Total	100% (7)	100% (26)	100% (12)	100% (8)	100% (32)

Data and Number have a quite even distribution of items across all three word-count groups, probably because it is easier to find authentic contexts for 15-year-olds for quantitative or statistical problems.

ITEM-FORMAT AND MATHEMATICS PERFORMANCE

There is research evidence that item format can influence students' performance in different countries (O'Leary, 2001) and that this can vary for different levels of ability (Routitsky and Turner, 2003). In this section the analyses investigate how item-format associates with item difficulty and whether there is an interaction between item format and other features of the items discussed earlier in this report. The question of whether there is a format by country interaction is also studied. Differences between countries can pose important questions about instructional and assessment practices in these countries. For this purpose, a full factorial analysis of variance is used in the same way as it was used in previous sections (see Annex A5 for these results).

How the question is asked can have an impact on its difficulty.

In the PISA 2003 initial report, mathematics items were represented by constructed response or selected response items.

Constructed response items can be subdivided into the following two categories (see Chapter 3 for the examples listed below):

- *Extended open constructed response*: response requires some explanation or justification of the answer referred to as the "extended response" type of "open constructed-response" items (see, for example, GROWING UP Q3).

- *Short answer*: response requires a number as an answer (see, for example, GROWING UP Q1 and EXPORTS Q1).

- *Multiple short answer*: response requires several numbers as an answer, and these answers were scored as one item (see, for example, SKATEBOARD Q3).

Selected response items can be subdivided into the following two categories:

- *Simple multiple choice items* (see, for example, COLORED CANDIES Q1).
- *Complex multiple choice items* (see, for example, CARPENTER Q1).

The analysis in this section is based on the categorical variable item-format with the five categories described above.

ITEM-FORMAT AND ITEM DIFFICULTY ACROSS COUNTRIES

The most difficult questions are extended response and complex multiple choice.

Figure 5.5 shows the average item difficulty for each item-format category in each country. It is not clear from this figure which item-format categories are significantly different from each other. Number of items, means and standard deviations are presented for each item-format category in Table 5.6.

Table 5.6 shows that on average the most difficult item type is *extended response* followed by *complex multiple choice*. The easiest item type is simple *multiple choice*. It also shows that the *extended response* type and the *multiple choice* type vary between countries more than the *complex multiple choice* type and the *short answer* type.

Figure 5.5 ■ **Average item difficulty (logits) by item-format by country**

Multiple comparisons for mean difficulties using Bonferroni adjustment (see Annex A5, Table A5.6) show that all item-format categories are significantly different from each other at the 0.01 level. However, analysis of variance with item-format and countries used as factors (see Annex A5, Table A5.1) shows that there is no interaction effect between countries and item-format.

There are a few countries for which the mean item difficulty in some of the item-format categories is more than two standard deviations away from the overall mean for these item types. For Brazilian students the multiple choice items appear to be relatively easier (-0.87 logits) while the short answer items appear to be relatively more difficult (0.00 logits). For the Russian Federation and the Slovak Republic the short answer items appear to be relatively easier (-0.31 and -0.29 logits respectively). For Serbia the complex multiple choice items appear to be relatively easier (0.22 logits) while the extended response items are relatively more difficult (1.03 logits). Finally, for Korea the complex multiple choice items are relatively more difficult (0.63 logits) and the multiple choice items are relatively easier (-0.75 logits). These differences might give some indication to the specialist in national assessment and curriculum where to look for strengths and weaknesses.

ITEM-FORMAT, THE THREE C'S AND WORD-COUNT

As it was the case for the word-count, item-format shows a small but significant interaction with competencies and context (see Annex A5, Table A5.2): the competencies are a much stronger factor than item-format while context is weaker. Interactions between topics and item-format will not be discussed due to the very small number of items in each cell (see Annex A5, Table A5.3).

Content is still a stronger predictor of difficulty than question format.

There is an interesting relationship between the item-format and the word-count. Although analysis of variance shows a strong interaction, this is mainly due to the fact that all complex multiple choice items have more than 50 words and therefore none of them belong to the category of "short" answer items.

Table 5.6
Mean and standard deviation of item difficulty in item-format categories across countries

Item format type	Number of items	Item type difficulty across countries Mean	(SD) in logits
Complex multiple choice	11	0.40	(0.08)
Multiple choice	18	-0.52	(0.11)
Short answer	37	-0.16	(0.06)
Multiple short answer	5	0.03	(0.13)
Extended response	14	0.76	(0.13)

Figure 5.6 shows that the multiple choice, short answer and extended response categories behave very similarly in relation to word-count.

Figure 5.6 also shows that the item-format is a stronger predictor of item difficulty than the word-count (see Annex A5, Table 5.4).

Figure 5.6 ■ **Average relative difficulty of questions by item-format and word-count**

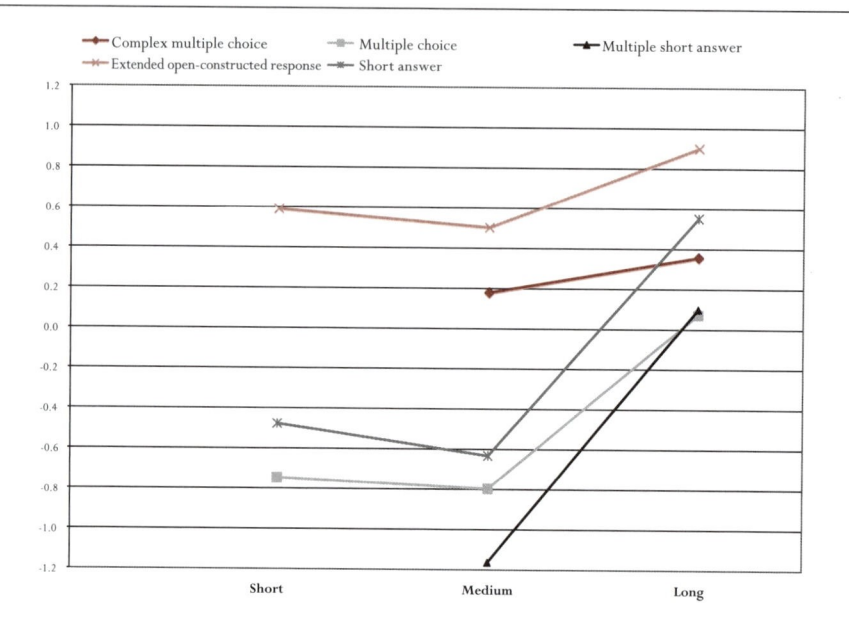

DIFFERENCES IN ITEM-FORMAT AND OMISSION RATES

A question's content, context, format, word count and difficulty are all related to whether a student attempts to answer it or not.

Another issue related to item-format is differences in omission rates to items. The responsiveness of students in terms of patterns of missing values can depend on the different surface characteristics of the test item as well as on its difficulty. These item characteristics include: format, content, context, reading demand and amount and complexity of information in the stimulus. The results of this section will have important implications for assessment practices and instruction.

While examining Michigan's High School Proficiency Test, DeMars (2000) found an interaction between test consequences (high/low stakes) and item-format. She also argues that "motivation and performance may be influenced by item response format". There is also a general belief that non-response is somewhat higher for constructed response items than for multiple-choice items, although it could be an effect of item difficulty (Lord, 1975; Dossey, Mullis, and Jones, 1993).

To investigate the relationship between the amount of missing data and item format, data from the PISA 2003 field trial were examined. These data were used because they were coded to better reflect the nature of the missing data. The percent of missing data was calculated for each item in each of the following item types: multiple choice, extended response, and short answer. For this analysis, the calculation of missing data identified and excluded non-reached items. In other words, missing data are here defined as "embedded missing" responses, and "trailing missing" responses are not included.

The correlation between the percentage of missing responses and item difficulty was calculated. This gives a measure of the degree to which missing data can be explained by item difficulty for each format type. For the multiple choice items the correlation was 0.331, for the extended response items the correlation was 0.499 and for the short answer items the correlation was 0.737.

The distribution of the omissioin rates by item-format shows that the amount of missing data in multiple choice items, which comprise the easiest set, varies from 1.66% to 17.62%. Here the relative item difficulty accounts for about 11% of the variation of missing data.

At the same time, the results show that the amount of missing data in extended response items, which comprise the most difficult set, varies from 9.78% to 57.54%. Here the relative item difficulty accounts for about 25% of the variation of missing data.

Finally, the results show that the amount of missing data in short answer items, which are slightly easier than extended response items, varied from 2% to 48%. Here the relative item difficulty accounts for about 54% of the missing data.

The scoring of student responses for PISA treats missing responses (excluding non-reached) as incorrect. This is based on the assumption that students omit an item because they do not know how to answer it. Such an assumption is supported more strongly when there is a strong relationship between item difficulty and omission rates. In this study, the short answer items fit this model best and the multiple choice items the least.

There also appear to be some other factors, other than item difficulty, particularly for the multiple choice and for the extended response items that contribute to the causes for missing data. There is a widespread belief that for multiple choice items, if students don't know the answer, they have the possibility of guessing, and therefore omission rates are low. This possibility does not exist for the extended response format types. What other factors might apply? Item difficulty is one possibility. Other particular factors that may contribute to missing data might include the reading load of the item. Further investigation is required.

To further investigate the relationship between missing data and item format type, the average amount of missing data for each item format type described above (multiple choice, extended response and short answer types) was calculated. The results are shown in Table 5.7. The results are reported separately for a sequence of item difficulty ranges to control for difficulty.

Table 5.7
Average percent of missing data by item difficulty for three item-format categories – PISA Field Trial 2003

Difficulty range (logits)	Average percent of missing data		
	Multiple Choice	**Short Answer**	**Extended Response**
Less than -2	2.90%	2.65%	N/A
Between -2 and -1	3.89%	8.98%	N/A
Between -1 and 0	4.96%	9.45%	17.80%
Between 0 and 1	6.30%	19.28%	21.44%
Between 1 and 2	9.12%	24.96%	29.00%
Between 2 and 3	6.82%	31.83%	33.62%
More than 3	N/A	28.35%	48.56%

This table shows that the percent of missing data in each difficulty range is the lowest for the multiple choice items and slightly lower for the short answer items than for the extended response items. In addition it shows that the general trend within each format type is that the more difficult the item, the more missing data is observed, but for the most difficult multiple choice items, the percent of missing data is lower than expected, which raises a question about guessing. Is there some difficulty threshold for the multiple choice items beyond which students will guess rather than omit the item?

Overall the PISA 2003 field trial data shows that while both item format type and item difficulty play significant roles in the amount of missing data, there are other features of items such as the length and complexity of the stimulus and the form in which choices are presented that also play a role.

CONCLUSION

This chapter examines the relationship between the difficulty of the PISA 2003 mathematics questions, and features such as the amount of reading required, the type of questions asked and the percentage of students who do not answer each question.

The correlation between question difficulty and the number of words in the question was weak because the question difficulty does not reflect small changes in the number of words used. The difficulty of the question is only influenced once the questions are of a certain length: on average there is no difference in difficulty between short questions (less than 50 words) and medium-length questions (between 50 and 100 words). Long questions (more than 100 words) were significantly more difficult across all countries. From this point of view, although words were counted in English and the meaning of "short", "medium" and "long" will be different in different languages, the three word-count categories were stable across all countries and, therefore, appropriate for analysis.

The analysis shows that there is a small but significant interaction between the number of words used in the questions and the context in which the question is presented, the mathematical content and the mathematical competencies required in answering the question. Further, the number of words used in the questions is a better predictor of question difficulty than the context in which the question is presented, but weaker than the mathematical content and the mathematical competencies required in answering the question. Another finding is that although on average short and medium-length questions are of the same difficulty, when subdivided by the context in which the question is presented, the mathematical content or the mathematical competencies, medium-length questions consistently show that they can be easier than short questions. One possible explanation for this finding is that as long as the stimulus and the question itself are not too wordy, some additional words, if appropriate, can help to solve the problem rather than make it more difficult.

Content and competency are stronger predictors of difficulty than context or word count.

In the PISA 2003 field trial data, a strong correlation was found between students' general performance in mathematics and their preferential performance on multiple choice items versus open-ended extended response items (see Routitsky and Turner, 2003). That is, lower ability students are performing better than expected on the multiple choice items and higher ability students are performing better than expected on the open-ended extended response items. This was partially explained by the combination of psychometric characteristics of these types of items (lower discrimination for multiple choice items) in the PISA 2003 field trial test and the wide range of students' abilities. This was avoided in the PISA 2003 Main Survey by keeping item discrimination in a narrower range. In this chapter the analysis is concentrated on the PISA 2003 Main Survey items. As with the word-count, the item-format shows a small but significant interaction with context and competencies. The three main item-format categories – multiple choice, extended open ended response and short answer – do not have an interaction with word-count. Generally, the item-format is a better predictor of item difficulty than the context and the word-count but weaker than the competencies. This means that format considerations should be treated with caution when tests are constructed, especially when students from a wide range of abilities are tested.

In relation to omission rates, the findings of this chapter show that while both the item-format and item-difficulty play significant roles in the amount of missing data, there are other features of items such as the length and complexity of the stimulus and form in which choices are presented that also play a role. Cultural differences among countries may explain a portion of differential non-response due to cultural views about guessing when one does not know the answer.

While format and difficulty are related to students not attempting to answer questions, word count and cultural bias also play a role.

In conclusion, the analyses carried out in this chapter show that PISA results can provide useful information about different features of questions, how these features relate to each other and how relevant they are to the difficulty of the questions. The most important factor influencing the difficulty of the PISA mathematics questions is the mathematical competencies or the cognitive demands required in answering the question, followed by the mathematical content (represented as traditional topics), the type of question (item-format), and the amount of reading required to understand the question (word-count). The factor showing the weakest influence over the difficulty of the PISA mathematics questions was the context in which the question is presented. These findings can be helpful in developing both mathematics assessments and mathematics text books, as well as for classroom teachers when making a choice of questions for instruction and assessment.

Content is the only factor related to difficulty that showed important variation across countries.

The only factor that showed differentiation by countries was the mathematical content of the questions (as represented by the traditional topic). For all other factors there were no significant variations between countries within each factor: context in which the question is presented, word-count, mathematical competencies required in answering the question.

Mathematical Problem Solving and Differences in Students' Understanding

This chapter concentrates on problem solving methods and differences in students' mathematical thinking. It discusses the processes involved in what is referred to as the "mathematisation" cycle. The chapter provides two case studies, explaining how the elements required in the different stages of mathematisation are implemented in PISA items.

INTRODUCTION

In problem-solving students apply their mathematical literacy using different methods and approaches.

PISA 2003 made a special effort to assess students' problem solving, as this is where *mathematical literacy* has its real application in life. The correlation between students' performance on overall mathematics items and their performance on those specifically focusing on problem solving was 0.89, which is higher than the correlation between mathematics and science (0.83). Nevertheless, analyses of assessment results on problem solving showed that students doing well in problem solving are not simply demonstrating strong mathematical competencies. In fact, in many countries students perform differently in these two domains (OECD, 2004b).

This chapter explains how mathematical problem-solving features are revealed in PISA questions. The PISA 2003 assessment framework (OECD, 2003) gives rise to further possibilities for investigating fundamentally important mathematical problem-solving methods and approaches. In particular, the framework discusses processes involved using the term mathematisation. The scoring design of PISA 2003 mathematics questions does not always allow for a full study of the patterns in students' responses in relation to their mathematical thinking; nevertheless, the discussion of the questions where the full problem-solving cycle comes alive can be useful for instructional practices.

PISA can also be used to analyse student strategies and misconceptions.

One area of the analysis of PISA items of particular interest to mathematics educators is the focus on student strategies and misconceptions. Misconceptions, or the study of students' patterns of faulty performances due to inadequate understandings of a concept or procedure, are well documented in the mathematics education literature (Schoenfeld, 1992; Karsenty, Arcavi and Hadas, 2007). Although PISA was not set up to measure misconceptions, the use of double scoring of some of the PISA items and the particular focus of others allow for findings of instructional interest to mathematics educators.

GENERAL FEATURES OF MATHEMATICAL PROBLEM SOLVING IN PISA

Mathematisation refers to the problem-solving process students use to answer questions.

The section begins with description of the "problem-solving process" or the process of "mathematisation" as it is called in the PISA framework of mathematical literacy (OECD, 2003). Two case studies of PISA questions that make the problem-solving cycle visible are then presented.

The mathematisation cycle …

The "problem-solving process" is generally described as a circular process with the following five main features:

1. Starting with a problem based in a real-world setting.

2. Organising it according to mathematical concepts and identifying the relevant mathematics.

3. Gradually trimming away the reality through processes such as making assumptions, generalising and formalising, which promote the mathematical

features of the situation and transform the real-world problem into a mathematical problem that faithfully represents the situation.

4. Solving the mathematical problem.

5. Making sense of the mathematical solution in terms of the real situation, including identifying the limitations of the solution.

Figure 6.1 shows the cyclic character of the mathematisation process.

The process of mathematisation starts with a problem situated in reality (1).

Figure 6.1 ■ **Mathematisation cycle**

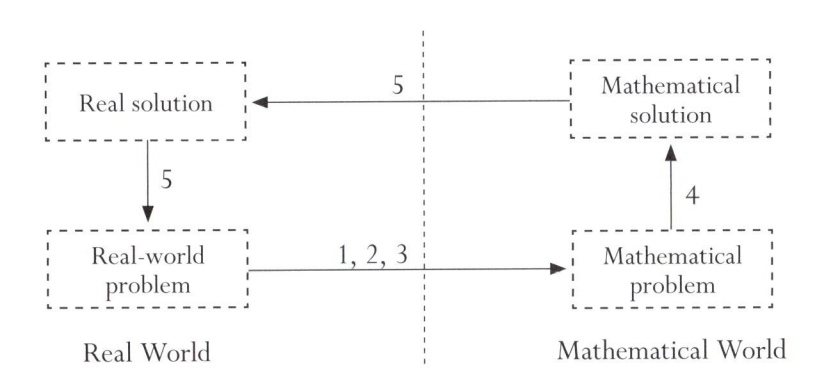

```
┌─────────────┐         5          ┌─────────────┐
│   Real      │◄───────────────────│ Mathematical│
│  solution   │                    │  solution   │
└─────────────┘                    └─────────────┘
      │ 5                                 ▲ 4
      ▼                                   │
┌─────────────┐                    ┌─────────────┐
│ Real-world  │     1, 2, 3        │ Mathematical│
│  problem    │───────────────────►│  problem    │
└─────────────┘                    └─────────────┘

    Real World                     Mathematical World
```

Next, the problem-solver tries to identify the relevant mathematics and reorganises the problem according to the mathematical concepts identified (2), followed by gradually trimming away the reality (3). These three steps lead the problem-solver from a real-world problem to a mathematical problem.

The fourth step may not come as a surprise: solving the mathematical problem (4).

Now the question arises: what is the meaning of this strictly mathematical solution in terms of the real world? (5)

These five aspects can be clustered into three phases according to general features of mathematical problem-solving approaches (see, for example, Polya, 1962; and Burkhardt, 1981):

… and the three phases of mathematisation.

Phase 1. Understanding the question (*e.g.* dealing with extraneous data), which is also called horizontal mathematisation.

Phase 2. Sophistication of problem-solving approaches, which is also referred to as vertical mathematisation.

Phase 3. Interpretation of mathematical results (linking mathematical answers to the context).

MAKING THE PROBLEM-SOLVING CYCLE VISIBLE THROUGH CASE STUDIES OF QUESTIONS

Two case studies of mathematisation in PISA questions.

There is some real-world mathematical problem-solving present in all PISA mathematics questions. However, not all of the PISA mathematics questions make the full cycle of problem-solving clearly visible due to the limited time that students have to answer the questions: the average allowable response time for each question is around two minutes, which is too short a period of time for students to go through the whole problem-solving cycle. The PISA mathematics questions often require students to undertake only part of the problem-solving cycle and sometimes the whole problem-solving cycle. This section presents two case studies of questions where students are required to undertake the full problem-solving cycle.

The first case study: Bookshelves – Question 1

BOOKSHELVES

Question 1: BOOKSHELVES

To complete one set of bookshelves a carpenter needs the following components:

- 4 long wooden panels,
- 6 short wooden panels,
- 12 small clips,
- 2 large clips and
- 14 screws.

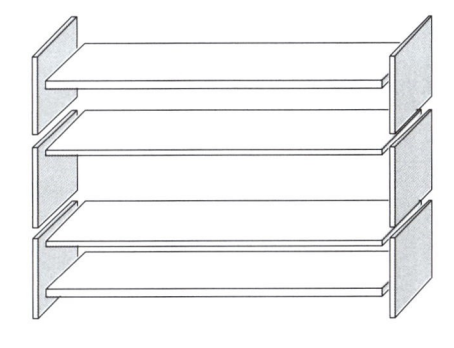

The carpenter has in stock 26 long wooden panels, 33 short wooden panels, 200 small clips, 20 large clips and 510 screws.

How many sets of bookshelves can the carpenter make?

Answer: ....

BOOKSHELVES SCORING QUESTION 1

Full Credit
Code 1: 5

No Credit
Code 0: Other responses

Code 9: Missing

STEP 1

The problem starts in a real-world context, and actually this reality is authentic. The problem is presented in a rather "natural" way, that is to say, there is a limited amount of text with a functional visual underpinning. However, the question is somewhat less complex than most problems are in reality due to the fact that there is almost no irrelevant or redundant information given in the question. This is important in light of step 2 of the problem-solving cycle: where students need to organise the facts in a more or less mathematical way and identify the relevant mathematics.

STEP 2

Many students will take a moment to check how the text stating the required components for a set of bookshelves relates to the picture. They will probably find out that this is not of much help since the only additional information is about how the long wooden panels relate to the short wooden panels.

So the relevant numbers are: one set requiring 4, 6, 12, 2 and 14; we have available 26, 33, 200, 20 and 510. These are the main components of step 2.

STEP 3

The students now translate the problem to the mathematical world. The question "how many sets can be completed before running out of one of the necessary components?" can be reformulated into a mathematical problem in the following way. Students need to look for the highest multiple of the first set (4, 6, 12, 2, 14) that fits into the other set (26, 33, 200, 20, 510).

STEP 4

In this step the students solve the problem.

One possibility for students that gives a high degree of confidence is producing the following table:

(4	6	12	2	14)	FOR 1 set
(8	12	24	4	28)	FOR 2 sets
(12	18	36	6	42)	FOR 3 sets
(16	24	48	8	56)	FOR 4 sets

Students list each row of components until they run out of one of the components:

| (20 | 30 | 60 | 10 | 70) | FOR 5 sets |
| (24 | 36 | 72 | 12 | 84) | FOR 6 sets |

Finally students run out of one component; there are only 33 short wooden panels available, and the last row to make a sixth set shows the need for 36 short wooden panels.

So, mathematically speaking, the highest multiple of

(4   6   12   2   14) that fits into (26   33   200   20   510) is 5.

It is very likely that students would use this strategy, but other strategies are also possible. Another possible strategy is to first identify the crucial component. If students really understand the problem right away from a more mathematical point of view, they might be tempted to calculate the ratios of the components: $^{24}/_4 = 6$ + remainder; $^{33}/_6 = 5$ + remainder; $^{200}/_{12}$; $^{20}/_2$; $^{210}/_{14}$ are abundant, because each of them is greater than 10, so the answer is 5.

For a fully correct answer, students simply need to answer "5". Such an answer allows no further insights into the processes of mathematical problem solving followed by students. The correct answer is reached without concluding the problem-solving cycle.

STEP 5

To complete the problem-solving cycle, students would need to make sense of the meaning of the solution ("5") in terms of the real world. That is quite obvious here: with the listed available components, only five complete sets of bookshelves can be made. However, it is also possible to identify that the critical component is the short wooden panels, and that with three more of these it is possible to produce six sets.

Reflection on Bookshelves – Question 1

Although this problem seems straightforward, the difficulties involved in solving it should not be underestimated. There is a particular risk that lower achieving students could skip step 5, skip the reflection on the answer, and give 6 as an answer. These students would most likely use the "ratio" strategy, find the ratio to be 5.5 (6 to 33) and not reflect properly on the meaning of this number.

Double-digit coding could also be used to collect data on the different strategies students use, and the results could answer questions like:

- did the students use the first strategy: "build until you run out"?

- did the students use the "ratio" or "most critical component" strategy?

- did the students use another strategy?

The use of double-digit coding and specific answers and data associated with questions such as these would give us a better understanding about the nature and level of *mathematical literacy* of the students.

BOOKSHELVES Q1 is an example of a PISA mathematics question that requires a rather simple mathematical problem-solving process because students seem to know quite well what the problem is all about and how to solve it in a mathematical way. However, the PISA mathematics assessment also includes questions where the mathematical problem solving is more challenging, as there is no known strategy available to the students. SKATEBOARD Q3 is a good example of such a question.

The second case study: Skateboard – Question 3

SKATEBOARD

Eric is a great skateboard fan. He visits a shop named SKATERS to check some prices.

At this shop you can buy a complete board. Or you can buy a deck, a set of 4 wheels, a set of 2 trucks and a set of hardware, and assemble your own board.

The prices for the shop's products are:

Product	Price in zeds	
Complete skateboard	82 or 84	
Deck	40, 60 or 65	
One set of 4 Wheels	14 or 36	
One set of 2 Trucks	16	
One set of hardware (bearings, rubber pads, bolts and nuts)	10 or 20	

Question 3: SKATEBOARD

Eric has 120 zeds to spend and wants to buy the most expensive skateboard he can afford.

How much money can Eric afford to spend on each of the 4 parts? Put your answer in the table below.

Part	Amount (zeds)
Deck	
Wheels	
Trucks	
Hardware	

SKATEBOARD SCORING QUESTION 3

Full Credit
Code 1: 65 zeds on a deck, 14 on wheels, 16 on trucks and 20 on hardware.

No Credit
Code 0: Other responses

Code 9: Missing

SKATEBOARD Q3 seems, at least at first glance, to have some similarities with BOOKSHELVES Q1. Students have to construct something, there are components, and both questions are presented in an authentic context. But mathematically speaking these questions are different, as the discussion will show.

STEP 1

The problem starts in a real-world context, and actually reflects a reality for many students in their daily life. For students who are unfamiliar with skateboards, photos are provided to give them some necessary information. SKATEBOARD Q3 is an example of a real situation for students as well. Students have a certain amount of money to spend and want to buy the best quality skateboard for their money.

STEP 2

It seems relatively straightforward for students to organise the problem. There are four components, and for three of the four, students need to make a choice (the only component for which there is no choice is the trucks). It is easy for students to identify the relevant mathematics since they have to add numbers and compare a sum with a given number.

A worksheet could look like:

Deck	40	60	65
4 Wheels	14	36	
Trucks	16		
Hardware	10	20	
TOTAL	120		

STEP 3

Mathematically speaking students have to find one number from each of the four categories that will result in the maximum sum within given restrictions. The restrictions for those numbers are: the first number has to be 40, 60 or 65; the second has to be 14 or 36; the third is 16; the fourth is either 10 or 20; and the sum cannot exceed 120. These are all the necessary elements to solve the problem.

STEP 4

Solving the mathematical problem is a little bit different than in BOOKSHELVES Q1 as there are no known strategies available to the students. This means that students will likely "fall back" on the trial-and-error method. This is actually a well known strategy, but every time it is applied, it is new within the context of the problem. Given the small amount of numbers that the students have to deal with, they can actually start making a list of all possibilities without running out of time. Given the task "to buy the most expensive", it seems advisable for students to start with the larger numbers from each collection:

65 + 36 + 16 + 20. These add up to 137, which is too much.

So students have to save 17 zeds. There are the following possibilities to save money:

On the deck:	5 or 25 zed
On the wheels:	22 zed
On the trucks:	nothing
On the hardware:	10 zed

This list makes the solution clear: save on the wheels, and students spend 115 zeds.

This strategy is structured. The problem with trial-and-error strategies lies often in the unstructured approach that students use. Students give different answers including:

40, 36, 16, 20

60, 14, 16, 20

60, 36, 16, 10

65, 36, 16, 20

The fact that students are not asked to give an explanation means that it is not possible to analyse their reasoning in more detail. A more detailed coding scheme like double-digit coding would allow for further insights into the use of actual strategies or reasoning and thinking.

STEP 5

This step was not tested in this question. It would be possible if students had been asked to explain their solutions. However, this question required students to fill out the numbers in a table. With appropriate argumentation, one of the solutions given above (40, 36, 16, 20) might be considered as a "better" solution. For example, the student who came up with this answer might say that having excellent quality wheels is much more important than having a better deck. This might be a very good argument, indeed, but without the argument it is impossible to know whether this was actually the student's reasoning.

Reflection on Skateboard – Question 3

The problem-solving cycle becomes apparent in almost all aspects in SKATEBOARD Q3. The problem-solving strategy most often used is not a routine procedure. However, it is not possible to shed light on the actual problem-solving process, because in the present format, and with the restrictions of many large-scale tests, the relevant information is not collected for identifying the thinking and argumentation processes used to solve these problems. If such a question were used in daily practices of instruction in schools, it would offer opportunities for discussion and argumentation. It is possible to ask additional questions, and in particular, to require students to give arguments for their solutions.

STUDENTS' MATHEMATICAL UNDERSTANDINGS AND ITEM SCORING

This section will analyse students' understandings of particular mathematics topic areas overall and by subgroups. The section begins by surveying the nature of the information available from the different types of PISA mathematics questions, offering examples of information about students' mathematical understandings. Three more case studies of PISA 2003 mathematics items are then presented. The first examines results from a collection of questions that relate to proportional reasoning, which is a very important core mathematical topic. The second looks at those questions that involve some symbolic algebra and the third looks at average (mean). It is clear that form does make a difference when referring to the item format in which questions are presented (Braswell and Kupin, 1993; Traub, 1993; Dossey, Mullis, and Jones, 1993).

Students' understanding of a topic varies across and within countries

Item coding in the database and information on students' thinking

The data available from each question are dependent on the type of coding used to identify responses to the question. PISA has a variety of question formats, and each of them have been coded in a certain way (Table 6.1). For some questions, the students' answers were entered directly into software (*e.g.* the distractor, circled by students in the multiple-choice question or a simple numeric answer in some of the questions requiring short answers). Sometimes for technical reasons these responses were later recoded (*e.g.* numeric items were recoded automatically as 1 if the answer was correct and 0 otherwise). For PISA 2003, both recoded (scored) and actual (raw) information on such questions is available on the international database.

Other questions were coded by qualified coders following the international coding guides. Some of these questions can only provide data on the number of correct and incorrect responses or auxiliary information. The auxiliary information covers missing and invalid responses and the number of students to whom the question was not administered. Others (so called double-digit questions) provide possible insights into students' strategies, errors, cognitive obstacles and misconceptions.

Table 6.1
Use of different types of PISA 2003 mathematics question formats

Coding type	Number of questions
Directly entered responses	
Multiple-choice questions	17
Complex multiple-choice questions	11
Numeric response	21
Coded Responses	
Single-digit (including partial credit)	27(2)
Double-digit (including partial credit)	9(7)

In this section, examples of some of the coding types are given, along with some discussion of how this can be used to identify students' thinking.

Simple scoring of correct or incorrect answers (single-digit full credit coding)

Some questions only provide data on whether responses are correct, incorrect or missing. For example, the response categories for EXPORTS Q1 were recorded only as correct (Code 1, OECD average of 74%), incorrect (Code 0, OECD average of 17%) or missing (Code 9, OECD average of 9%).

The coding of PISA questions provides clues to the student's understanding.

With this coding, the difficulty of these questions can be compared with that of other questions, both in terms of the percentage of students giving an incorrect answer and the percentage of students who do not attempt to answer the question. However, no information is available on the methods, correct or incorrect, used by students. The PISA study has adopted strict protocols to decide which constructed responses should be regarded as correct. In the case of the very easy EXPORTS Q1, students need to identify the height of the column (27.1) that is associated with 1998. Responses were marked correct if they answered 27.1 (without a unit), 27.1 million zeds, or 27 100 000 zeds. Even with such a simple question and simple response type, the criteria for correctness can make an important difference in student success.

EXPORTS

The graphics below show information about exports from Zedland, a country that uses zeds as its currency.

Total annual exports from Zedland in millions of zeds, 1996-2000

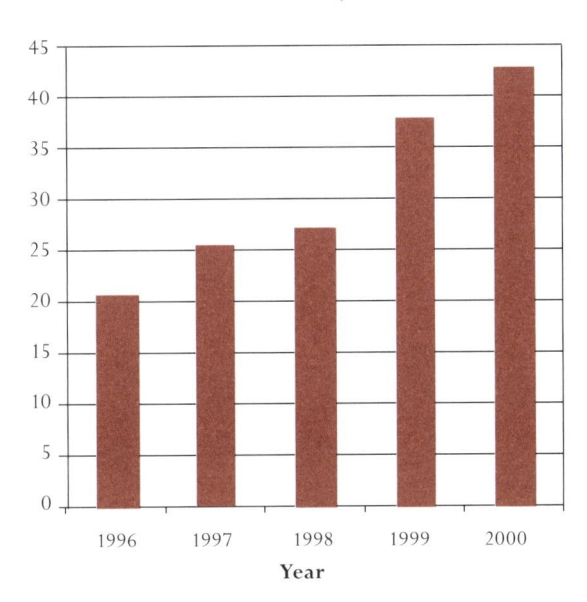

Distribution of exports from Zedland in 2000

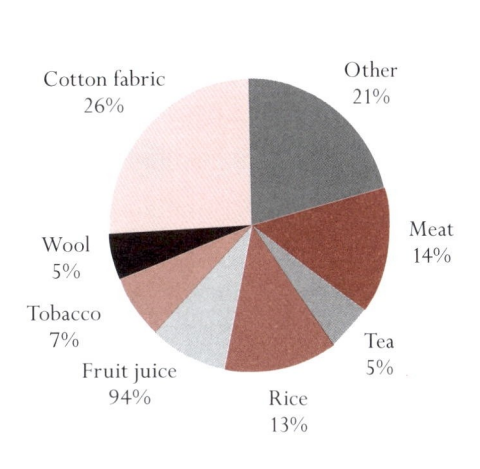

- Cotton fabric 26%
- Other 21%
- Meat 14%
- Tea 5%
- Rice 13%
- Fruit juice 94%
- Tobacco 7%
- Wool 5%

Question 1: EXPORTS

What was the total value (in millions of zeds) of exports from Zedland in 1998?

Answer: ....

EXPORTS SCORING QUESTION 1

Full Credit
Code 1: 27.1 million zeds or 27 100 000 zeds or 27.1 (unit not required)

No Credit
Code 0: Other responses

Code 9: Missing

Multiple-choice questions

Multiple-choice questions often provide more information, because the methods that students are likely to have used can sometimes be inferred from the choices (correct and incorrect) that were made. So, for example, EXPORTS Q2 is a multiple-choice question. Table 6.3 lists other examples.

Question 2: EXPORTS

What was the value of fruit juice exported from Zedland in 2000?

- A 1.8 million zeds.
- B 2.3 million zeds.
- C 2.4 million zeds.
- D 3.4 million zeds.
- E 3.8 million zeds.

EXPORTS SCORING QUESTION 2

Full Credit
Code 1: E. 3.8 million zeds

No Credit
Code 0: Other responses

Code 9: Missing

The distractors for EXPORTS Q2 are all calculated as approximately 9% of a quantity indicated on the bar graph (the total annual exports from 1996 to 2000 *i.e.* 9% of 20.4, 9% of 25.4, etc.).

Distractor C is the most frequent response after the correct response (E) both for the OECD average (Table 6.2) and also for a very large majority of countries. This is probably because the year 1998 was involved in the previous question EXPORTS Q1. This shows how student performance on a question can often be affected by irrelevant aspects of the question, the type or presentation of the question, or the students' failure to read all of the required information prior to answering.

These particular multiple-choice distractors have all been constructed in the same way, as 9% of a quantity on the column graph. This assumes that students find from the pie graph that 9% is associated with fruit juice, know that they need to find 9% of a quantity from the column graph, can calculate 9% of the quantity, but have difficulty identifying the correct quantity from the column graph.

Table 6.2
Distribution of responses for Exports – Question 2

	Distractor	Percent of students selecting the distractor (OECD average)	How the answer can be calculated
A	1.8 million zeds.	11%	9% of 1996 data = 1.836
B	2.3 million zeds.	10%	9% of 1997 data = 2.286
C	2.4 million zeds.	16%	9% of 1998 data = 2.439
D	3.4 million zeds.	8%	9% of 1999 data = 3.411
E	3.8 million zeds.	48%	9% of 2000 data = 3.834
	Missing response	7%	

Of itself, the coding provides no information about whether this is indeed an accurate description of the solution paths used by a majority of students. Students may have estimated 9% by eye (correctly or incorrectly), or they may have selected something a little less than 10% of 42.6 million (correctly or incorrectly).

Where students have not used a correct calculator or paper-and-pencil algorithm for finding the percentage, it is likely that they have calculated an answer not given amongst the multiple-choice distractors. For example, a common error in finding 9% would be to divide by 9, rather than to multiply, and this would lead to an answer greater than any of the distractors. Other potential obstacles for students that are not probed by the set of distractors are related to the use of millions. The problem would be easier for students who work directly in millions of zeds at every stage than for those students who convert to 42 600 000 zeds (or an incorrect version) and then have to convert back to the given form of the answer. This is likely to result in order of magnitude errors, which are also not tapped by these distractors. Students making calculation errors that result in answers not amongst the supplied distractors might omit the question, select the nearest, or try the calculation again.

Table 6.3
Examples of multiple-choice questions in Chapter 3

Question	Where to find question in Chapter 3
EXPORTS – Question 2	*Examples of easy questions section*
SKATEBOARD – Question 2	
EARTHQUAKE – Question 1	
COLOURED CANDIES – Question 1	*Examples of questions of moderate difficulty section*
NUMBER CUBES – Question 2[1]	
CARPENTER – Question 1[1]	*Examples of difficult questions section*

1. These questions are classified as complex multiple-choice questions as they require students to provide a series of correct answers from predefined choices.

Simple scoring of correct, incorrect or partially correct answers (single digit partial credit coding)

WALKING Q1 provides an example of a different coding pattern for responses called partial credit coding.

WALKING

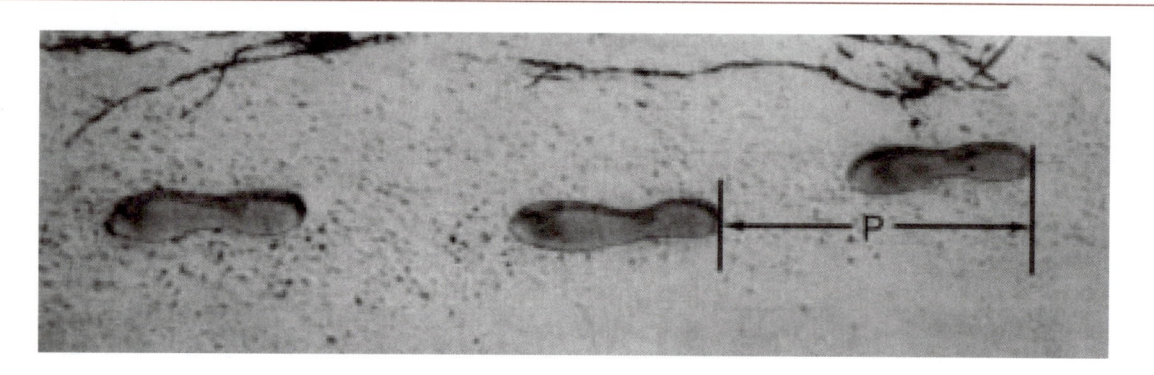

The picture shows the footprints of a man walking. The pacelength P is the distance between the rear of two consecutive footprints.

For men, the formula, $\frac{n}{P} = 140$, gives an approximate relationship between n and P where

   n = number of steps per minute, and

   P = pacelength in metres.

Question 1: WALKING

If the formula applies to Heiko's walking and Heiko takes 70 steps per minute, what is Heiko's pacelength? Show your work.

WALKING SCORING QUESTION 1

Full Credit

Code 2: 0.5 m or 50 cm, $\frac{1}{2}$ (unit not required)

$$\frac{70}{P}$$

70 = 140 P.

P = 0.5.

$$\frac{70}{140}$$

Partial Credit
Code 1: Correct substitution of numbers in the formula, but incorrect answer, or no answer.

$\dfrac{70}{P} = 140$ [substitute numbers in the formula only].

$\dfrac{70}{P} = 140$.

$70 = 140\,P$.

$P = 2$ [correct substitution, but working out is incorrect].

OR

Correctly manipulated the formula into $P = \dfrac{n}{140}$, but no further correct work.

No Credit
Code 0: Other responses
70 cm.

Code 9: Missing

A correct response to this question is obtained by substituting the number of paces per minute ($n = 70$) into the formula $n/P = 140$ and finding P, the length of a pace. The maximum score of 2 is given to responses of 0.5 m, 50 cm or ½ m, with or without units. The partial credit score of 1 identifies a group of students with various incomplete algebra understandings: those who can substitute $n = 70$ into the formula, but do not solve the resulting equation correctly for P; those who can rearrange the formula to make the unknown pace length (P) the subject without going further; and those who obtain the likely incorrect answer $P = 2$. Other responses score 0. The OECD average on this question is 36% for the score 2, 22% for the score 1 and, 21% for the score 0 with 21% missing responses. In the PISA protocols of 2003, partial credit coding generally provides information on how much progress students have made towards a solution, rather than the type of progress that they have made.

Complex scoring recording the type of methods used in correct and partially correct answers (double-digit coding)

WALKING Q3 provides an example of "double-digit" coding (see Table 6.4 for further examples and the Chapter 3 section on examples of difficult questions in the PISA 2003 mathematics assessment). As with WALKING Q1 partial credit can be awarded in conjunction with "double-digit" coding. The double digits provide more information about the student's response and potentially about the method that they used and errors that they made (see Dossey, Jones and Martin, 2002).

Question 3: WALKING

Bernard knows his pacelength is 0.80 metres. The formula applies to Bernard's walking.

Calculate Bernard's walking speed in metres per minute and in kilometres per hour. Show your working out.

WALKING SCORING QUESTION 3

Full Credit
Code 31: Correct answers (unit not required) for both metres/minute and km/hour:
  N = 140 × 0.80 = 112.
  Per minute he walks 112 × 0.80 metres = 89.6 metres.
  His speed is 89.6 metres per minute.
  So his speed is 5.38 or 5.4 km/hr.

  Code 31 as long as both correct answers are given (89.6 and 5.4), whether working out is shown or not. Note that errors due to rounding are acceptable. For example, 90 metres per minutes and 5.3 km/hr (89 × 6) are acceptable.
  89.6, 5.4.
  90, 5,376 km/h.
  89.8, 5376 m/hour [note that if the second answer is given without units, it should be coded as 22].

Partial Credit (2 points)
Code 21: As for Code 31 but fails to multiply by 0.80 to convert from steps per minute to metres per minute. For example, his speed is 112 metres per minute and 6.72 km/hr.
  112, 6.72 km/h.

Code 22: The speed in metres per minute correct (89.6 metres per minute) but conversion to kilometres per hour incorrect or missing.
  89.6 metres/minute, 8960 km/hr.
  89.6, 5376.
  89.6, 53.76.
  89.6, 0.087 km/h.
  89.6, 1.49 km/h.

Code 23: Correct method (explicitly shown) with minor calculation error(s) not covered by Code 21 and Code 22. No answers correct.
  n = 140 × 0.8 = 1120; 1120 × 0.8 = 896. He walks 896 m/min, 53.76 km/h.
  n = 140 × 0.8 = 116; 116 × 0.8 = 92.8. 92.8 m/min → 5.57 km/h.

Code 24: Only 5.4 km/hr is given, but not 89.6 metres/minute (intermediate calculations not shown.

WALKING Q3 is complex. From the given value of *P*, the pace length, students have to use the formula to work out *n*, how many paces per minute, multiply *P* by *n* to get the distance travelled per minute and then convert this to kilometres per hour. As partially correct answers are credited for this question, there are several possible scores: 3 for a fully correct answer; 2 for a high-level partially correct answer; 1 for a low-level partially correct answer, or 0 for an incorrect answer. However, information is also separately recorded about the methods used by students and other features of their solutions. In this case, the following errors are separately coded for students scoring 2 points:

- finding *n* correctly (steps per minute) but not multiplying by *P* to get metres per minute, but working correctly in all other respects (code 21);

- incorrect conversions of the correct metres per minute speed to km/hr (code 22);

- minor calculation errors in a fully correct method (code 23);

- correct km/hr speed without supporting calculations (code 24).

WALKING Q3 was a difficult question. The OECD average of students scoring 3 (fully correct) was 8% and the total of students falling into all codes of the high-level partially correct answer (score 2) was 9%. Across the OECD countries, the most common reasons why students did not get the answer fully correct (score 3) were a failure to calculate metres per minute (4.8% of students on average) or an incorrect conversion to kilometres per hour (3.3% of students on average). Only a minority of students made minor calculations errors (0.58% of students on average) or gave the correct kilometres per hour speed without supporting calculations (0.22% of students on average).

The countries with high percentages of students giving fully correct answers to this question (score 3) also tend to have high percentages of students who just fall short of giving fully correct answers (score 2): in fact there is a strong correlation (0.86 across the OECD countries) between these two groups of students. For example, the only countries with more than 16% of students giving fully correct answers were Japan (18%) and the partner economy Hong Kong-China (19%). These countries, and other countries with relatively high percentages of students scoring 3, had the highest percentages of students scoring 2. Only in Japan (14%) and the partner economies Hong Kong-China (22%) and Macao-China (18%) were there more than 10% of students in code 21 for WALKING Q3 and only in the partner economies Hong Kong-China (4%) and Macao-China (4%) were there more than 0.5% of students in code 24. For code 22, the partner country Liechtenstein had the highest percentage of students (8%) and Liechtenstein had the fourth highest percentage of students scoring on this question overall (32% correct). Five to six percent of students scored in code 22 in the Czech Republic, the Slovak Republic, France, Poland, Hungary and the partner country the Russian Federation and this was the most common reason why students did not score fully correct answers in these countries.

Countries with similar performance patterns over the PISA mathematics assessment also show similarities in the frequency of double-digit codes. For example, it was noted in Chapter 3 that the partner economies Hong Kong-China and Macao-China have similar performance patterns across PISA mathematics questions. They also show similar patterns of responses within questions using the double-digit codes.

The aim of PISA is to measure *mathematical literacy*, therefore the use of double-digit coding generally indicates whether students have used a correct method to solve the mathematical problem, but have small calculation errors, rather than to show the nature of the errors that students make or to measure the incidence of established misconceptions and common error patterns.

Double coding can help disentangle student's problem-solving strategies and understanding.

The above summary of types of coding shows that each question format can provide interesting data for analysis in relation to students' approaches to mathematical problems. Simple multiple-choice questions with one correct answer among a set of well-constructed distractors can provide valuable information about the prevalence of misconceptions, at least as they are restricted to students' selecting them from a list of alternatives. Items calling for students to construct and provide a simple numeric response provide slightly better information in that students have to show what they would do without being prompted by a set of distractors. The analysis still provides correctness and distribution data.

At yet a higher level, single-digit coded items with partial credit offer a deeper insight into students' performances. In many countries, such partial credit scoring shows that many students start correctly, move to an intermediate result, and then fail to complete the task set by the problem in the unit. Partial credit shows that they have command of subsidiary knowledge and skill, even while students did not attend to the full task presented. Finally, double-digit coding provides mathematics educators with a picture of the degree of correctness, including partial credit, as well as a picture of the relative distribution of strategies employed by students. All coding types provide information about difficulty of items, but additional information shows the differing depths of understanding that exist within and between countries.

Table 6.4
Examples of questions with double-digit coding in Chapter 3

Question	Where to find question in Chapter 3
GROWING UP – Question 2	*Examples of the easiest questions section*
GROWING UP – Question 3	*Examples of questions of moderate difficulty section*
EXCHANGE RATE – Question 3	
WALKING – Question 3	*Examples of difficult questions section*
ROBBERIES – Question 1	

STUDENTS' UNDERSTANDING OF PROPORTIONAL REASONING

A wide range of important mathematical ideas with strong application in the real world involve proportional reasoning. Proportional reasoning is an important part of the "conceptual field of multiplicative structures" (Vergnaud 1983, 1988), which consists of situations where multiplication and division are usually required, including problems involving ratios, rates, scales, unit conversions, direct variation, fractions, and percentages as well as simple multiplication and division problems involving discrete objects or measures. It takes a very long time for students to master multiplicative structures. Additive structures (involving addition and subtraction) are principally developed in the early grades, but understanding of multiplicative structures continues to develop from the early years of schooling through to the PISA age groups. A critical part of the learning in central content areas is the differentiated roles played by conceptual content and procedural content and the building of connections between "knowing what" and "knowing how" (Hiebert, 1986).

The prevalence of proportional reasoning in PISA questions

Evidence of the importance of proportional reasoning to a mathematically literate person comes from the observation that 11 of the 85 questions in the PISA 2003 mathematics assessment involved proportional reasoning, and this contributed to many others. For example, EXPORTS Q2 requires knowledge of percentages, and the unit EXCHANGE RATE requires conversion of currencies. Table 6.5 gives some other examples. Ability in proportional reasoning underlies success in many

Table 6.5
Instances of proportional reasoning in questions presented in Chapter 3

Unit name	Questions (see Chapter 3)	Skills required	Percent correct (OECD average)
WALKING	Question 3	Includes conversion of metres per minute to kilometres per hour (minor part of question) with other skills.	8%
GROWING UP	Question 3	Involves concept of rate of growth, with other skills.	45%
ROBBERIES	Question 1	An absolute difference is to be judged as a relative difference.	14%
EXCHANGE RATE		Converting currencies by	
	Question 1	multiplying by 4.2	80%
	Question 2	dividing by 4.0 and	74%
	Question 3	explaining which rate is better.	40%
EXPORTS	Question 2	Finding 9% of 42.6 million, along with graph reading skills.	46%
COLOURED CANDIES	Question 1	Finding a probability (relative frequency), with simple graph reading skills.	50%
STAIRCASE	Question 1	Apportioning a rise of 252 cm over 14 stairs.	78%

aspects of the school curriculum so teachers need to pay particular attention to students' understandings of this important area. There are underlying proportional reasoning abilities, as well as specific knowledge required to deal with its many occurrences (such as probability, percentage, speed, slope, rate of change, *etc.*).

The difficulty of proportional reasoning questions

Within the mathematics education literature, major factors that contribute to the difficulty of proportional reasoning have been identified. Some important factors are the nature of the real context and the numerical nature of the ratio, rate, or proportion involved. For example, students deal with simple ratios much more easily, both procedurally and conceptually, than with complex ratios. It is much easier to identify that a problem can be solved by multiplying a quantity by 3 than that it can be solved by multiplying by $11/12$. When the rate, ratio or proportion is a number between 0 and 1, students often choose division when multiplication is appropriate, and vice versa.

One simple and useful classification of levels of proportional reasoning questions is provided by Hart (1981). The Hart scale provides a guide to classifying the difficulty of proportional reasoning questions. The description of the Hart levels is given in Table 6.6 and selected PISA questions that have a major emphasis on proportional reasoning, with question difficulties and PISA proficiency levels are given in Table 6.7.

The data in Table 6.7 indicate that the proportional reasoning questions show a wide range of difficulty, from -1.85 to 3.21, although the difficulty of the hardest item, the unreleased THE THERMOMETER CRICKET Q2, is increased by the use of algebra. The Hart scale and the PISA scale basically follow a similar pattern with two exceptions (EXCHANGE RATE Q3 and POPULATION PYRAMIDS Q3 to be discussed below). Classification of PISA questions testing proportional reasoning according to the Hart (1981) scale shows that the majority of these questions could be classified as having a difficulty of two or three. Hart considered that students did not really display proportional reasoning until they were able to be successful on level 3 items. Since the PISA difficulty of the Hart level 2 items is around 0, this indicates that the development of proportional reasoning ability must still be considered an important instructional goal for teachers of many 15-year-old students.

Table 6.6
Hierarchy of proportional reasoning items (Hart, 1981)

Level	Description and examples
1	Problems involving doubling, trebling or halving. Rate given.
2	Rate easy to find or given. Also, problems which can be solved by adding appropriate quantity and half the quantity. Example: to change a recipe for 4 to a recipe for 6, take the original recipe and add half of each quantity.
3	Rate is more difficult to find. Also, when fraction operations are involved. Example: Mr Short is 4 matchsticks tall or 6 paperclips. Mr Tall is 6 matchsticks tall; how tall in paperclips?
4	Use of ratio or rate needs to be identified. Questions complex in numbers in ratio or in setting. Example: making an enlargement in the ratio of 5:3.

Table 6.7
Level of difficulty of proportional reasoning questions
(PISA proficiency level, question difficulty parameter, Hart level)

PISA question	PISA proficiency level	Question difficulty parameter	Hart level
EXCHANGE RATE – Question 1	1	-1.85	1
EXCHANGE RATE – Question 2	2	-1.36	2
BICYCLES – Question 2	2	-1.20	2
CHAIR LIFT – Question 1	4	0.04	2
EXPORTS – Question 2	4	0.14	3
CHAIR LIFT – Question 2	4	0.22	3
EXCHANGE RATE – Question 3	4	0.45	2
CARBON DIOXIDE – Question 3	5	0.93	3
BICYCLES – Question 3	5	1.55	4
POPULATION PYRAMIDS – Question 3	6	1.71	3
THE THERMOMETER CRICKET – Question 2	6	3.21	4

The results of the PISA proportional reasoning questions showed wide variation across countries, but groups of countries which have been shown to perform similarly in other studies once again performed similarly. For example, Figure 6.2 shows that the pattern of performance of a selection of the English-speaking countries is very similar. The items are arranged in rank order of question difficulty.

Figure 6.2 ■ **Performance of some English speaking countries on proportional reasoning items, illustrating their similar pattern of performance**

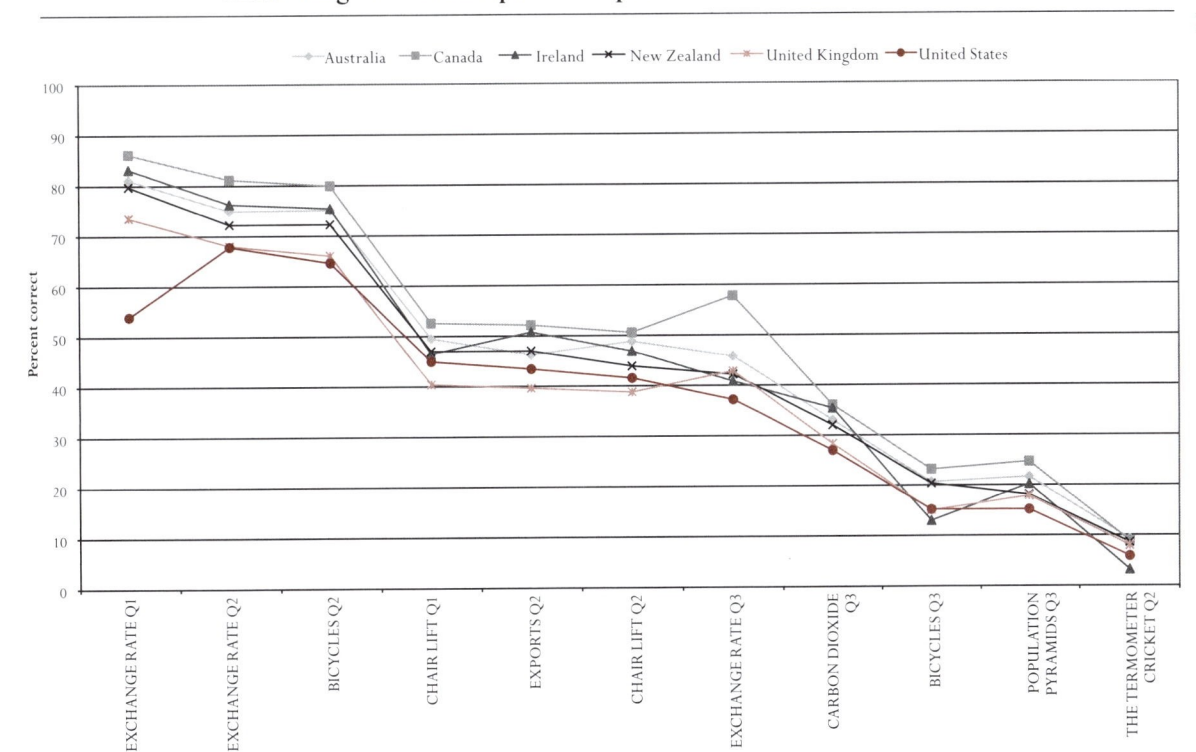

EXCHANGE RATE Q3, which is seen to behave unusually in Table 6.11, shows an interesting variation.

EXCHANGE RATE

Mei-Ling from Singapore was preparing to go to South Africa for 3 months as an exchange student. She needed to change some Singapore dollars (SGD) into South African rand (ZAR).

Question 1: EXCHANGE RATE

Mei-Ling found out that the exchange rate between Singapore dollars and South African rand was:

1 SGD = 4.2 ZAR

Mei-Ling changed 3000 Singapore dollars into South African rand at this exchange rate.

How much money in South African rand did Mei-Ling get?

Answer: ....

EXCHANGE RATE SCORING QUESTION 1

Full Credit
Code 1: 12 600 ZAR (unit not required)

No Credit
Code 0: Other responses
Code 9: Missing

Question 2: EXCHANGE RATE

On returning to Singapore after 3 months, Mei-Ling had 3 900 ZAR left. She changed this back to Singapore dollars, noting that the exchange rate had changed to:

1 SGD = 4.0 ZAR

How much money in Singapore dollars did Mei-Ling get?

Answer: ....

EXCHANGE RATE SCORING QUESTION 2

Full Credit
Code 1: 975 SGD (unit not required)

No Credit
Code 0: Other responses
Code 9: Missing

Question 3: EXCHANGE RATE

During these 3 months the exchange rate had changed from 4.2 to 4.0 ZAR per SGD.

Was it in Mei-Ling's favour that the exchange rate now was 4.0 ZAR instead of 4.2 ZAR, when she changed her South African rand back to Singapore dollars? Give an explanation to support your answer.

EXCHANGE RATE SCORING QUESTION 3

Full Credit
Code 11: Yes, with adequate explanation.

- Yes, by the lower exchange rate (for 1 SGD) Mei-Ling will get more Singapore dollars for her South African rand.
- Yes, 4.2 ZAR for one dollar would have resulted in 929 ZAR. [Note: student wrote ZAR instead of SGD, but clearly the correct calculation and comparison have been carried out and this error can be ignored]
- Yes, because she received 4.2 ZAR for 1 SGD and now she has to pay only 4.0 ZAR to get 1 SGD.
- Yes, because it is 0.2 ZAR cheaper for every SGD.
- Yes, because when you divide by 4.2 the outcome is smaller than when you divide by 4.
- Yes, it was in her favour because if it didn't go down she would have got about $50 less.

No Credit
Code 01: Yes, with no explanation or with inadequate explanation.

- Yes, a lower exchange rate is better.
- Yes, it was in Mei-Ling's favour, because if the ZAR goes down, then she will have more money to exchange into SGD.
- Yes, it was in Mei-Ling's favour.

Code 02: Other responses.

Code 99: Missing.

Figure 6.3 shows examples of two pairs of otherwise similar countries Austria and Sweden, performing differently on this question. Whereas the first two questions in the EXCHANGE RATE unit required calculations, the third question, probed understanding more conceptually (see Chapter 4).

The proportional reasoning questions show a small gender difference. On average, the percent correct for males is 2% greater than for females. This is independent of the question difficulty. This is an unexpected finding since it is often the case that gender differences are most pronounced with high performance.

These looks into proportional reasoning provide mathematics educators and curriculum specialists with comparative understanding of 15-year-olds' understanding of aspects of proportionality. The links to Hart's theoretical model

Figure 6.3 ■ **Proportional reasoning performances of Austria and Sweden, showing variation in Exchange Rate – Question 3**

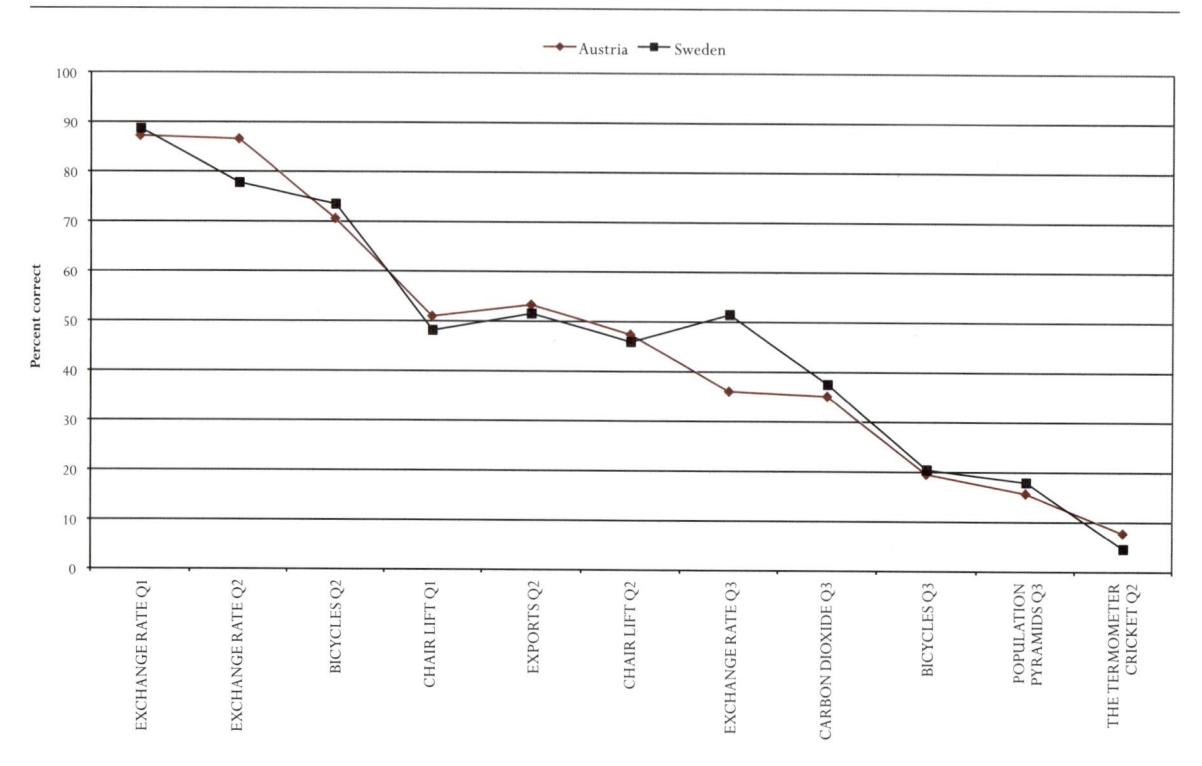

and the correlation of the PISA difficulty findings with the hierarchical levels of the model indicates that PISA findings can be used as one data source for the possible validation of such theoretical models. But, beyond that, the data on proportionality serves as an international look at differences in the deeper understandings students have, country-by-country, in a mathematical area directly underlying the study of linear equations.

STUDENTS' UNDERSTANDING OF SYMBOLIC ALGEBRA

This section looks at student performance on the questions explicitly using symbolic algebra. There are very few such questions, which reflect the on-going debates in the mathematics education community about the place of symbolic algebra in the school curriculum. This topic has been greatly affected by the "massification" of schooling, and the lack of obvious relevance of algebraic symbols in students' everyday lives has meant that there has been substantial questioning of its role in many countries, often resulting in substantial adjustment to the algebra curriculum.

The PISA questions succeed in finding contexts and problems that are properly part of *mathematical literacy*, and show that even symbolic algebra has a place in *mathematical literacy*. At the same time, the low number of such questions within the PISA 2003 mathematics assessment indicates that dealing with algebraic symbols is not of extremely high importance for *mathematical literacy*

(MacGregor and Stacey, 1997; Stacey and MacGregor, 2000). Out of seven PISA questions classified as Algebra (see Annex A3) six questions use algebraic symbols explicitly, while the other uses only the graphical representation. This question is the non-released Q2 of the CONTAINERS unit.

In the PISA framework, the *change and relationships* overarching idea is rightly recognised to be much larger than the set of problems where algebraic letter symbols are used. Indeed it is also larger than algebra, as it includes work about patterns, functions and variation that can be represented graphically, numerically and spatially as well as symbolically.

Letter symbols appear in PISA in formulas, which is one of their several uses within and outside mathematics. Letters in a formula are generally relatively easy to conceptualise because they have a clear referent (a quantity to be used in calculation). Moreover, the purpose of a formula is clear: to calculate one quantity from others. For example, in the formula $A = LW$, the letter A stands for the area of a rectangle, and L and W stand for side lengths. The purpose of the formula is to calculate the area (A) from its length (L) and width (W). Happily, the characteristics that make formulas a relatively easy part of symbolic algebra also make them a very important part of early algebra.

The graphs in Figure 6.4 and Figure 6.5 show the performance of selected countries on the six questions that used letter symbols in a formula. The unit WALKING contains two of them (see Chapter 3). OECD average percent

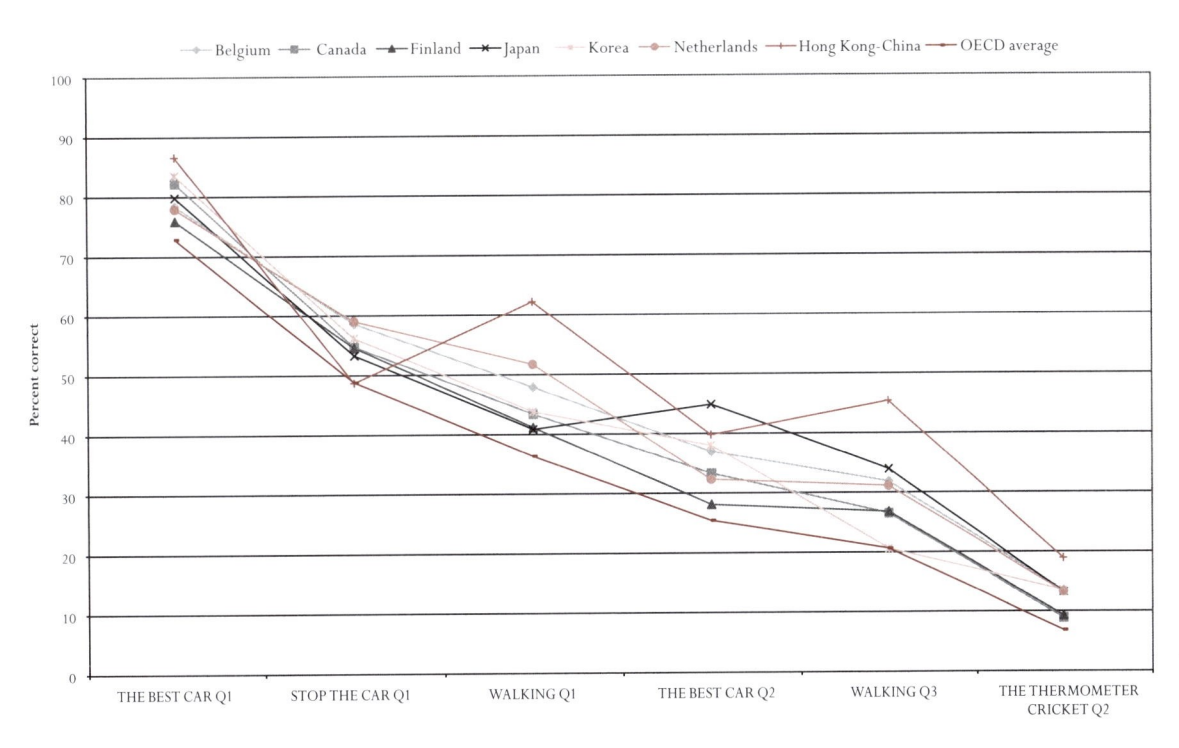

Figure 6.4 ■ **Performance on algebra items for countries scoring highly on the content items from *change and relationships***

Figure 6.5 ■ **Performance on algebra items for the countries scoring at the OECD average on the content items from *change and relationships***

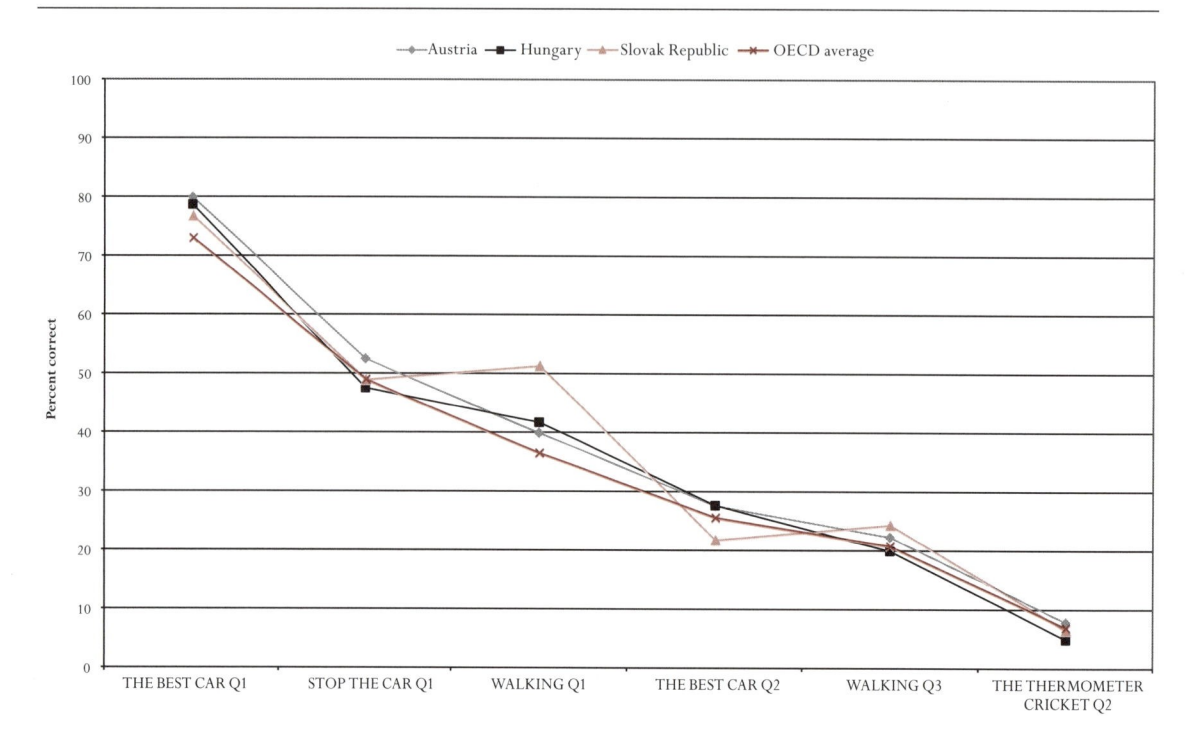

correct is 35% for WALKING Q1 and 20% for the more difficult WALKING Q3. Figure 6.4 shows the performance of the best performing countries in the *change and relationships* overarching area and Figure 6.5 shows the performance of a group of countries not statistically different from the OECD average.

These six questions show a wide range of performance. Examination of the use of algebraic symbols in each question shows little link to the algebraic demand of the questions (substituting, interpreting, using and writing formulas). For example, the most complicated algebraic expressions are in the second easiest question (non-released question STOP THE CAR Q1). However, it is the case that writing a formula from worded information was required only in the most difficult (non-released question THE THERMOMETER CRICKET Q2). As noted above, this question also had a high proportional reasoning demand, so the writing of the formula does not entirely explain the difficulty. The variation in performance on WALKING Q1 and WALKING Q3 (see Chapter 3), which is evident in both Figures 6.4 and 6.5, may indicate that students had difficulty interpreting this question, especially in some languages. Some evidence for this is that the western English-speaking countries do not change rank order on this question, as the countries in Figure 6.4 and Figure 6.5 do.

None of these six questions have double-digit coding that relates directly to students' understanding of algebra or the common algebraic errors.

STUDENTS' UNDERSTANDING OF AVERAGE

The following Figure 6.6 and Figure 6.7 show the percent correct for selected high-performing countries and the OECD average, on several non-released items that involve different types of knowledge about the average of a data set.

Within the real-world context of the non-released PISA unit HEIGHT, HEIGHT Q1 required an explanation of the procedure of calculating an average, HEIGHT Q2 tested knowledge of some of the properties of average, and HEIGHT Q3 required a difficult calculation of average.

Figure 6.6 shows that these countries have similarly high success rates on the item that tests knowledge of how to calculate an average from data, all above the OECD average. In fact, the variability of the country averages on this item is considerably larger than on the other items, but the selection of countries hides this fact. The central item of the figure, HEIGHT Q2, shows that the percentage of students with good knowledge of the properties of averages from these countries is spread, with the Netherlands having the highest percent correct, and some countries that were high performing on the previous item having less than the OECD percent correct. The third item, HEIGHT Q3, requires a difficult calculation of average, which can be made easier by a good conceptual understanding, as has been tested in the previous item. Here the graph shows that Korea, with relatively poor conceptual understanding performs extremely well. This is an interesting result, which is also evident in the results of Japan, although the effect is less marked.

Figure 6.6 ■ **Results of selected countries on HEIGHT concerning the mathematical concept of average**[1]

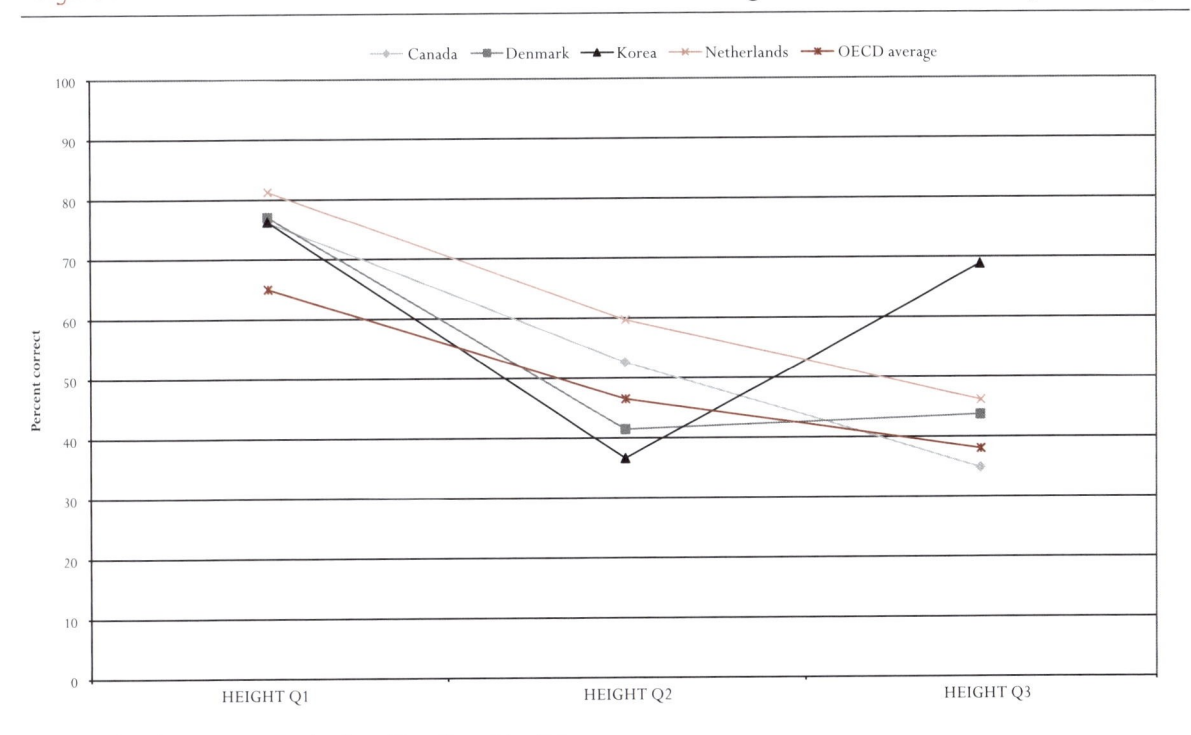

1. This is the percentage scoring 3 or 4 out of 4, on this partial credit item.

It seems that for some of the countries displayed, the calculation of the difficult average has been a challenging problem, where students have had to marshal their own mathematical problem solving resources. Good conceptual knowledge of average may have helped to solve this item. On the other hand, Korea shows a contrasting effect. Here there are the same percent correct on the difficult and straightforward items. This indicates that Korean students are able to treat this problem solving item as a routine task.

The three items in Figure 6.7 also show a progression. HEIGHT Q1, the first item in the graph (and the previous graph), requires the explanation for calculating an average. The selected countries nearly retain this advantage to the second item of Figure 6.7, which requires the calculation of an average embedded in a more complex situation. The additional complexity has caused an average drop of about 13% in percent correct. The third item again requires interpretation related to the concept of average and its calculation. The Netherlands and Canada again have a high percent correct, as they did for item HEIGHT Q2 in Figure 6.9 and Denmark again shows relatively less conceptual knowledge of the meaning of average. However, in this instance, just as many Korean students can interpret the real situation as can carry out the complicated calculation of average. The interpretation in this item tends towards interpreting an effect on a calculation, rather than linking average to the original data set.

This may indicate that in the Korea, instruction focuses less on the meaning of average in terms of the original data set and more on the attributes of the formula.

Figure 6.7 ■ **Results of selected countries on some non-released items concerning the mathematical concept of average**

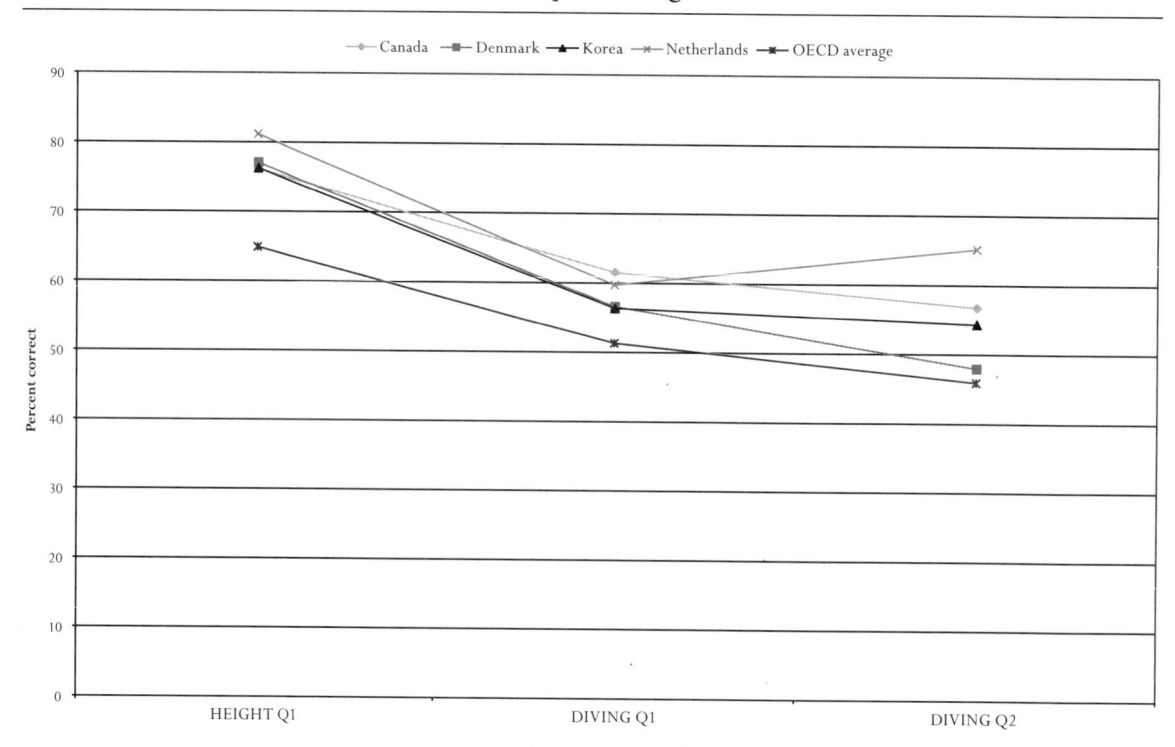

CONCLUSION

The PISA Framework of *mathematical literacy* (OECD, 2003) described the problem-solving process in terms of the process of "mathematisation". This chapter presented two examples of PISA items that make the problem-solving cycle visible. In these examples the problem-solving cycle comes alive in almost all aspects of the questions. Each problem solving strategy was not a routine procedure.

Unfortunately the authors cannot shed light on the actual specific strategy students used. The PISA scoring format does not provide specific information on the thinking and argumentation processes students actually used in solving problems.

At the level of daily practices of instruction in schools, however, it is possible to ask additional questions and, particularly, let students give arguments for their solutions. Teachers and other researchers might try using PISA items with their students and compare their results with those observed in this chapter.

Despite these limitations, the PISA database contains some insights on the problem-solving processes students use to tackle problems. First and foremost, the mathematical content and the selected contexts of the questions provide important examples of how mathematics is likely to be used in everyday life and work. Many items, for example, directly or indirectly involve proportional reasoning, which stresses the importance of this topic for *mathematical literacy*. The very large number of graphs in PISA questions also reflects the importance of this topic to modern living. Teachers might ask their students to relate problem structures presented in their classes to contexts from their personal lives to better connect the mathematics to students' self perceived needs and experiences.

When examining the PISA database through the lens of students' thinking about an individual mathematical topic, there is often little relevant data. The codes used in PISA aim principally to provide a good picture of mathematics in use. If there are good data for other purposes, it is serendipitous. Looking through the lens of a single mathematical topic, there may not be many PISA items to analyse. Furthermore, the coding may not record the common errors and any effect of a particular approach to the mathematical topic is likely to be masked by the mix of skills that is characteristic of mathematics being used in a context. In the future, to better reveal aspects of students' intra-mathematical thinking, it may be possible to sharpen the distractors used in multiple choice items and to include double digit coding. However, this should not be done at the expense of PISA primary goals.

REFERENCES

American Mathematical Society (AMS) (2009), *Mathematics on the Web: Mathematics by Classification*, retrieved 25 April 2009, from www.ams.org/mathweb/mi-mathbyclass.html.

Anderberg, M.R. (1973), *Cluster Analysis for Applications*, Academic Press, Inc., New York, NY.

Braswell, J. and **J. Kupin** (1993), "Item formats for assessment in mathematics", in R.E. Bennet and W.C. Ward (eds.), *Construction versus Choice in Cognitive Measurement: Issues in Constructed Response, Performance Testing, and Portfolio Assessment*, Lawrence Erlbaum, Hillside, NJ., pp. 167-182.

Burkhardt, H. (1981), *The Real World and Mathematics*. Blackie, Glasgow, Scotland.

Cohen, P.K. (2001), "Democracy and the numerate citizen: Quantitative literacy in historical perspective", in L.A. Steen (ed.), *Mathematics and Democracy: The Case for Quantitative Literacy*, National Council on Education and the Disciplines, Princeton, NJ., pp. 7-20.

de Lange, J. (2007), "Large-scale assessment and mathematics education", in F.K. Lester (ed.), *Second Handbook of Research on Mathematics Teaching and Learning*, Information Age Publishing, Charlotte, N.C., pp. 1111-1142.

DeMars, C.E. (2000), "Test stakes and item format interactions", *Applied Measurement in Education*, Vol. 13, No. 1, pp. 55-77.

Devlin, K. (1997), *Mathematics: The Science of Patterns*, Scientific American Library, New York, NY.

Dossey, J.A., I.V.S. Mullis and **C.O. Jones** (1993), *Can Students Do Mathematical Problem Solving?: Results from Constructed-Response Questions in NAEP's 1992 Mathematics Assessment*, National Center for Education Statistics, Washington DC.

Dossey, J.A., C.O. Jones and **T.S. Martin** (2002), "Analyzing student responses in mathematics using two-digit rubrics", in D.F. Robitaille and A.E. Beaton (eds.), *Secondary Analysis of the TIMSS Data*, Kluwer Academic Publishers, Dordrecht, The Netherlands.

Ellerton, N.F. and **M.A. Clements** (1991), *Mathematics in Language: A Review of Language Factors in Mathematics Learning*, Deakin University Press, Geelong, Victoria.

Fey, J.T. (1990), "Quantity", in L.A. Steen (ed.), *On the Shoulders of Giants: New Approaches to Numeracy*, National Academy Press, Washington DC., pp. 61-94.

Freudenthal, H. (1973), *Mathematics as an Educational Task*, D. Reidel, Dordrecht, The Netherlands.

Hart, K.M. (1981), *Children's' Understanding of Mathematics*, John Murray, London, UK.

Hiebert, J. (ed.) (1986), *Conceptual and Procedural Knowledge: The Case of Mathematics*, Erlbaum, Hillsdale, NJ.

Jakwerth, P.M., F.B. Stancavage and **E.D. Reed** (1999), *An Investigation of Why Students Do Not Respond to Questions*, NAEP Validity Studies, Working Paper Series, American Institutes for Research, Palo Alto, CA.

Karsenty, R., A. Arcavi and **N. Hadas** (2007), "Exploring informal mathematical products of low achievers at the secondary school level", *The Journal of Mathematical Behavior*, Vol. 26, No. 2, pp. 156-177.

Linn, R.L., E.L. Baker and **S.B. Dunbar** (1991), "Complex, performance-based assessment: expectations and validation criteria", *Educational Researcher*, Vol. 20, No. 8, pp. 15-21.

Lord, F.M. (1975), "Formula scoring and number-right scoring", *Journal of Educational Measurement*, Vol. 12, No. 1, pp. 7-11.

MacGregor, M. and **K. Stacey** (1997), "Students' understanding of algebraic notation: 11-16", *Educational Studies in Mathematics*, Vol. 33, No. 1, pp. 1-19.

Moore, D.S. (1990), "Uncertainty", in L.A. Steen (ed.), *On the Shoulders of Giants: New Approaches to Numeracy*, National Academy Press, Washington DC., pp. 95-137.

Mullis, I.V.S., M.O. Martin, E.J. Gonzalez, and **S.J. Chrostowski** (2004), *TIMSS 2003 International Mathematics Report: Findings From IEA's Trends in International Mathematics Science Study at the Fourth and Eighth Grades*, Boston University International TIMSS Study Center, Chestnut Hill, MA.

Nielsen, A.C., Jr. (1986), "Statistics in Marketing", in G. Easton, H.V. Roberts and G.C. Tiao (eds.), *Making Statistics More Effective in Schools of Business*, University of Chicago Graduate School of Business, Chicago, IL.

Organisation for Economic Co-operation and Development (OECD) (1999), *Measuring Student Knowledge and Skills: A New Framework for Assessment*, OECD, Paris.

OECD (2001), *Knowledge and Skills for Life: First Results from PISA 2000*, OECD, Paris.

OECD (2002), *PISA 2000 Technical Report*, OECD, Paris.

OECD (2003), *The PISA 2003 Assessment Framework: Mathematics, Reading, Science and Problem Solving Knowledge and Skills*, OECD, Paris.

OECD (2004a), *Learning for Tomorrow's World: First Results from PISA 2003*, OECD, Paris.

OECD (2004b), *Problem Solving for Tomorrow's World: First Measures of Cross-Curricular Competencies from PISA 2003*, OECD, Paris.

OECD (2005), *PISA 2003 Technical Report*, OECD, Paris.

OECD (2006), *Assessing Scientific Reading and Mathematical Literacy: A framework for PISA 2006*, OECD, Paris.

OECD (2009a), *PISA 2006 Technical Report*, OECD, Paris.

OECD (2009b), *Green at Fifteen? How 15-year-olds perform in environmental science and geoscience in PISA 2006*, OECD, Paris.

OECD (2009c), *Top of the Class – High Performers in Science in PISA 2006*, OECD, Paris.

OECD (2009d), *Equally prepared for life? How 15-year-old boys and girls perform in school*, OECD, Paris.

O'Leary, M. (2001), "Item format as a factor affecting the relative standing of countries in the Third International Mathematics and Science Study (TIMSS)", paper presented at the Annual Meeting of the American Educational Research Association, Seattle, WA., pp. 10-14.

Polya, G. (1962), *Mathematical Discovery: On Understanding, Learning and Teaching Problem Solving*, Wiley, New York, NY.

Routitsky, A. and **S. Zammit** (2002), "Association between intended and attained algebra curriculum in TIMSS 1998/1999 for ten countries", *Proceedings of the 2002 Annual Conference of the Australian Association for Research in Education*, retrieved 10 February 2010 from www.aare.edu.au/indexpap.htm.

Routitsky, A. and **R. Turner** (2003), "Item format types and their influence on cross-national comparisons of student performance", presentation given to the Annual Meeting of the American Educational Research Association (AERA) in Chicago, US, April.

Rutherford, A. (2001), *Introducing ANOVA and ANCOVA: A GLM Approach*, Sage Publications, Thousand Oaks, CA.

Schoenfeld, A. H. (1992). "Learning to think mathematically: Problem solving, metacognition, and sense making in mathematics", in D.A. Grouws (Ed.). *Handbook of Research on Mathematics Teaching and Learning*, Macmillan, New York, NY, pp. 334-370.

Senechal, M. (1990), "Shape", in L.A. Steen (ed.), *On the Shoulders of Giants: New Approaches to Numeracy*, National Academy Press, Washington DC., pp. 139-181.

Stacey, K. and **M. MacGregor,** (2000), "Learning the algebraic method of solving problems", *Journal of Mathematical Behavior*, Vol. 18, No. 2, pp. 149-167.

Steen, L.A. (ed.) (1990), *On the Shoulders of Giants: New Approaches to Numeracy*, National Academy Press, Washington D.C.

Stewart, K. (1990), "Change", in L.A. Steen (ed.), *On the Shoulders of Giants: New Approaches to Numeracy*, National Academy Press, Washington DC., pp. 183-217.

Thissen, D., and **H. Wainer** (2001). *Test Scoring*, Lawrence Erlbaum Associates, Mahwah, NJ.

Tout, D. (2001), "What is numeracy? What is mathematics?", in G.E. FitzSimons, J. O'Donoghue and D. Coben, (eds.), *Adult and Lifelong Education in Mathematics. A Research Forum. Papers from WGA 6 ICME 9*, Language Australia, Melbourne, pp. 233-243.

Traub, R.E. (1993), "On the equivalence of the traits assessed by multiple-choice and constructed-response tests", in R.E. Bennet and W.C. Ward (eds.), *Construction versus Choice in Cognitive Measurement: Issues in Constructed Response, Performance Testing, and Portfolio Assessment*, Lawrence Erlbaum, Hillside, NJ, pp. 29-44.

Vergnaud, G. (1983), "Multiplicative structures", in R. Lesh and M. Landau (eds.), *Acquisition of Mathematics Concepts and Processes*, Academic Press, New York, NY, pp. 127-174.

Vergnaud, G. (1988), "Multiplicative structures," In J. Hiebert and M. Behr (eds.), *Number Concepts and Operations in the Middle Grades*, National Council of Teachers of Mathematics, Reston, VA, pp. 141-161.

Wu, M.L., R.J. Adams and **M.R. Wilson** (1998), *ACER ConQuest: Generalised Item Response Modelling Software*, Melbourne, ACER Press.

Wu, M.L. (2006), "A comparison of mathematics performance between east and west – What PISA and TIMSS can tell us", in F.K.S. Leung, K.D. Graf and F.J. Lopez-Real (eds.), *Mathematics Education in Different Cultural Traditions: A Comparative Study of East Asia and the West,* ICMI Study 13**,** Springer Science and Business Media, New York, NY, pp. 239-259.

Zabulionis, A. (2001), "Similarity of mathematics and science achievement of various nations", *Education Policy Analysis Archives*, Vol. 9, No. 33, Education Policy Studies Laboratory, Arizona State University, Arizona, retrieved 10 February 2010, *epaa.asu.edu/epaa/v9n33*.

Annex A1

PISA 2003 MATHEMATICS ASSESSMENTS: CHARACTERISTICS OF QUESTIONS USED

Note that all the publicly released questions that were used in the PISA 2003 mathematics assessment are presented in Chapter 3. Note as well that the data in these graphs are from the compendium (available at *pisa2003.acer.edu.au/downloads.php*) and that the percentages refer to students that reached the question (the percentage of students who did not reach the question plus the percentage of students that attempted to answer it will therefore add up to more than 100%).

Annex A1

Table A1.1
Characteristics of released PISA 2003 mathematics items

Item code	Item name	OECD average percent correct	Item	Full 1	Partial 2	Partial 3	Traditional topic	Content area 'Overarching Idea'	Competency cluster	Context	Word count	Item format
M033Q01	P2000 A View Room Q1	76.77	432	432			Geometry	Space and shape	Reproduction	Personal	Short	Multiple Choice
M034Q01	P2000 Bricks Q1	43.27	582	582			Geometry	Space and shape	Connections	Educational and occupational	Medium	Short Answer
M124Q01	P2000 Walking Q1	36.34	611	611			Algebra	Change and relationships	Reproduction	Personal	Medium	Extended Response
M124Q03	P2000 Walking Q3	20.62	665	605	666	723	Algebra	Change and relationships	Connections	Personal	Medium	Extended Response
M144Q01	P2000 Cube Painting Q1	62.09	497	497			Geometry	Space and shape	Reproduction	Educational and occupational	Short	Short Answer
M144Q02	P2000 Cube Painting Q2	27.44	645	645			Geometry	Space and shape	Connections	Educational and occupational	Short	Short Answer
M144Q03	P2000 Cube Painting Q3	75.16	432	432			Geometry	Space and shape	Connections	Educational and occupational	Short	Multiple Choice
M144Q04	P2000 Cube Painting Q4	38.42	599	599			Geometry	Space and shape	Connections	Educational and occupational	Short	Short Answer
M145Q01	P2000 Cubes Q1	68.03	478	478			Data	Space and shape	Reproduction	Educational and occupational	Medium	Short Answer
M150Q01	P2000 Growing Up Q1	66.96	477	477			Number	Change and relationships	Reproduction	Scientific	Short	Short Answer
M150Q02	P2000 Growing Up Q2	68.77	472	420	525		Data	Change and relationships	Reproduction	Scientific	Medium	Short Answer
M155Q01	P2000 Population Pyramids Q1	64.86	479	479			Data	Change and relationships	Connections	Scientific	Medium	Short Answer

Learning Mathematics for Life: A Perspective from PISA – © OECD 2009

Table A1.1
Characteristics of released PISA 2003 mathematics items *(continued)*

Item code	Item name	OECD average percent correct	Item	Full/partial credit points 1	2	3	Traditional topic	Content area 'Overarching Idea'	Competency cluster	Context	Word count	Item format
M155Q02	P2000 Population Pyramids Q2	60.66	511	492	531		Data	Change and relationships	Connections	Scientific	Long	Extended Response
M155Q03	P2000 Population Pyramids Q3	16.79	674	643	706		Data	Change and relationships	Reflection	Scientific	Long	Extended Response
M179Q01	P2000 Robberies Q1	29.5	635	577	694		Data	Uncertainty	Connections	Public	Short	Extended Response
M192Q01	P2000 Containers Q1	40.41	594	594			Algebra	Change and relationships	Connections	Educational and occupational	Medium	Complex Multiple Choice
M266Q01	P2000 Carpenter Q1	19.95	687	687			Measurement	Space and shape	Connections	Educational and occupational	Medium	Complex Multiple Choice
M273Q01	P2000 Pipelines Q1	54.92	525	525			Geometry	Space and shape	Connections	Educational and occupational	Medium	Complex Multiple Choice
M302Q01	Car Drive Q1	95.32	262	262			Data	Change and relationships	Reproduction	Public	Medium	Short Answer
M302Q02	Car Drive Q2	78.42	414	414			Data	Change and relationships	Connections	Public	Short	Short Answer
M302Q03	Car Drive Q3	30	631	631			Data	Change and relationships	Reflection	Public	Long	Extended Response
M402Q01	Internet Relay Chat Q1	53.72	533	533			Measurement	Change and relationships	Connections	Personal	Medium	Short Answer
M402Q02	Internet Relay Chat Q2	28.79	636	636			Measurement	Change and relationships	Reflection	Personal	Long	Short Answer
M406Q01	Running Tracks Q1	28.66	639	639			Measurement	Space and shape	Connections	Public	Medium	Short Answer
M406Q02	Running Tracks Q2	19.33	687	687			Measurement	Space and shape	Connections	Public	Medium	Short Answer

Annex A1

Table A1.1
Characteristics of released PISA 2003 mathematics items *(continued)*

Item code	Item name	OECD average percent correct	Item	Full 1	Partial 2	Partial 3	Traditional topic	Content area 'Overarching Idea'	Competency cluster	Context	Word count	Item format
M406Q03	Running Tracks Q3	18.72	692	692			Measurement	Space and shape	Reflection	Public	Short	Extended Response
M408Q01	Lotteries Q1	41.6	587	587			Data	Uncertainty	Connections	Public	Long	Complex Multiple Choice
M411Q01	Diving Q1	51.39	545	545			Number	Quantity	Reproduction	Public	Long	Short Answer
M411Q02	Diving Q2	45.99	564	564			Number	Uncertainty	Connections	Public	Medium	Multiple Choice
M413Q01	Exchange Rate Q1	79.66	406	406			Number	Quantity	Reproduction	Public	Medium	Short Answer
M413Q02	Exchange Rate Q2	73.86	439	439			Number	Quantity	Reproduction	Public	Short	Short Answer
M413Q03	Exchange Rate Q3	40.34	586	586			Number	Quantity	Reflection	Public	Medium	Extended Response
M420Q01	Transport Q1	49.87	544	544			Data	Uncertainty	Reflection	Personal	Long	Complex Multiple Choice
M421Q01	Height Q1	64.97	485	485			Data	Uncertainty	Reproduction	Educational and occupational	Short	Extended Response
M421Q02	Height Q2	17.85	703	703			Number	Uncertainty	Reflection	Educational and occupational	Long	Complex Multiple Choice
M421Q03	Height Q3	38.04	602	602			Data	Uncertainty	Reflection	Educational and occupational	Short	Multiple Choice
M423Q01	Tossing Coins Q1	81.66	407	407			Data	Uncertainty	Reproduction	Personal	Medium	Multiple Choice
M434Q01	Room Numbers Q1*)			Deleted			Number	Quantity	Connections	Public	Long	Short Answer
M438Q01	Exports Q1	78.69	427	427			Data	Uncertainty	Reproduction	Public	Medium	Short Answer
M438Q02	Exports Q2	48.33	565	565			Number	Uncertainty	Connections	Public	Medium	Multiple Choice
M442Q02	Braille Q2	41.78	578	578			Number	Quantity	Reflection	Public	Long	Short Answer

196 Learning Mathematics for Life: A Perspective from PISA – © OECD 2009

Table A1.1
Characteristics of released PISA 2003 mathematics items *(continued)*

Item code	Item name	OECD average percent correct	Item	Full/partial credit points 1	2	3	Traditional topic	Content area 'Overarching Idea'	Competency cluster	Context	Word count	Item format
M446Q01	Thermometer Cricket Q1	68.22	470	470			Number	Change and relationships	Reproduction	Scientific	Medium	Short Answer
M446Q02	Thermometer Cricket Q2	6.79	801	801			Algebra	Change and relationships	Reflection	Scientific	Long	Short Answer
M447Q01	Tile Arrangement Q1	70.23	461	461			Geometry	Space and shape	Reproduction	Public	Short	Multiple Choice
M462Q01	Third Side Q1	14.11	702	671	734		Geometry	Space and shape	Reflection	Scientific	Short	Short Answer
M464Q01	The Fence Q1	25.11	662	662			Measurement	Space and shape	Connections	Public	Short	Short Answer
M467Q01	Coloured Candies Q1	50.21	549	549			Data	Uncertainty	Reproduction	Personal	Short	Multiple Choice
M468Q01	Science Tests Q1	46.77	556	556			Data	Uncertainty	Reproduction	Educational and occupational	Medium	Short Answer
M474Q01	Running Time Q1	74.07	452	452			Number	Quantity	Reproduction	Educational and occupational	Short	Multiple Choice
M484Q01	Bookshelves Q1	60.88	499	499			Number	Quantity	Connections	Educational and occupational	Medium	Short Answer
M496Q01	Cash Withdrawal Q1	53.12	533	533			Number	Quantity	Connections	Public	Medium	Complex Multiple Choice
M496Q02	Cash Withdrawal Q2	65.65	479	479			Number	Quantity	Connections	Public	Short	Short Answer
M505Q01	Litter Q1	51.55	551	551			Data	Uncertainty	Reflection	Scientific	Medium	Extended Response
M509Q01	Earthquake Q1	46.48	557	557			Data	Uncertainty	Reflection	Scientific	Long	Multiple Choice
M510Q01	Choices Q1	48.76	559	559			Number	Quantity	Connections	Educational and occupational	Medium	Short Answer
M513Q01	Test Scores Q1	32.21	620	620			Data	Uncertainty	Connections	Educational and occupational	Long	Extended Response

Annex A1

Annex A1

Table A1.1
Characteristics of released PISA 2003 mathematics items *(continued)*

Item code	Item name	OECD average percent correct	Item	Location on PISA scale (PISA score points) Full/partial credit points 1	2	3	Traditional topic	Content area 'Overarching Idea'	Competency cluster	Context	Word count	Item format
M520Q01	Skateboard Q1	72.01	480	464	496		Number	Quantity	Reproduction	Personal	Long	Short Answer
M520Q02	Skateboard Q2	45.53	570	570			Number	Quantity	Reproduction	Personal	Short	Multiple Choice
M520Q03	Skateboard Q3	49.78	554	554			Number	Quantity	Connections	Personal	Long	Short Answer
M547Q01	Staircase Q1	78.04	421	421			Number	Space and shape	Reproduction	Educational and occupational	Short	Short Answer
M555Q02	Number Cubes Q2	62.97	503	503			Geometry	Space and shape	Connections	Personal	Long	Complex Multiple Choice
M559Q01	Telephone Rates Q1	61	504	504			Number	Quantity	Reflection	Public	Long	Multiple Choice
M564Q01	Chair Lift Q1	49.26	551	551			Number	Quantity	Reproduction	Public	Medium	Multiple Choice
M564Q02	Chair Lift Q2	45.56	569	569			Number	Uncertainty	Reflection	Public	Medium	Multiple Choice
M571Q01	Stop The Car Q1	48.83	556	556			Algebra	Change and relationships	Reflection	Scientific	Long	Multiple Choice
M598Q01	Making a Booklet Q1	64.15	488	488			Geometry	Space and shape	Reflection	Personal	Long	Short Answer
M603Q01	Number Check Q1	47.1	559	559			Number	Quantity	Connections	Scientific	Long	Complex Multiple Choice
M603Q02	Number Check Q2	36.08	601	601			Number	Quantity	Connections	Scientific	Long	Short Answer
M702Q01	Support for President Q1	35.66	615	615			Data	Uncertainty	Connections	Public	Long	Extended Response
M704Q01	The Best Car Q1	72.91	447	447			Algebra	Change and relationships	Reproduction	Public	Long	Short Answer
M704Q02	The Best Car Q2	25.42	657	657			Algebra	Change and relationships	Reflection	Public	Long	Short Answer
M710Q01	Forecast of Rain Q1	33.88	620	620			Data	Uncertainty	Connections	Public	Long	Multiple Choice

Learning Mathematics for Life: A Perspective from PISA – © OECD 2009

Table A1.1
Characteristics of released PISA 2003 mathematics items *(continued)*

Item code	Item name	OECD average percent correct	Item	Full/partial credit points 1	2	3	Traditional topic	Content area 'Overarching Idea'	Competency cluster	Context	Word count	Item format
M800Q01	Computer Game Q1	91.77	309	309			Number	Quantity	Reproduction	Personal	Medium	Multiple Choice
M803Q01	Labels Q1	28.14	641	641			Number	Uncertainty	Connections	Educational and occupational	Medium	Short Answer
M806Q01	Step Pattern Q1	66.19	484	484			Number	Quantity	Reproduction	Educational and occupational	Short	Short Answer
M810Q01	Bicycles Q1	68.31	473	473			Number	Quantity	Connections	Personal	Medium	Short Answer
M810Q02	Bicycles Q2	71.71	459	459			Number	Quantity	Connections	Personal	Medium	Short Answer
M810Q03	Bicycles Q3	20.14	670	631	710		Number	Change and relationships	Reflection	Personal	Long	Extended Response
M828Q01	Carbon Dioxide Q1	39.74	593	593			Data	Change and relationships	Reproduction	Scientific	Long	Short Answer
M828Q02	Carbon Dioxide Q2	54.26	533	533			Data	Uncertainty	Connections	Scientific	Medium	Short Answer
M828Q03	Carbon Dioxide Q3	32.08	629	629			Number	Quantity	Connections	Scientific	Long	Short Answer
M833Q01	Seeing the tower Q1	31.81	628	628			Geometry	Space and shape	Connections	Personal	Long	Complex Multiple Choice

Annex A1

Annex A1

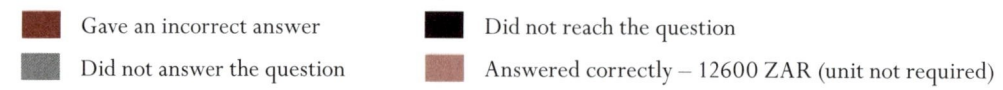

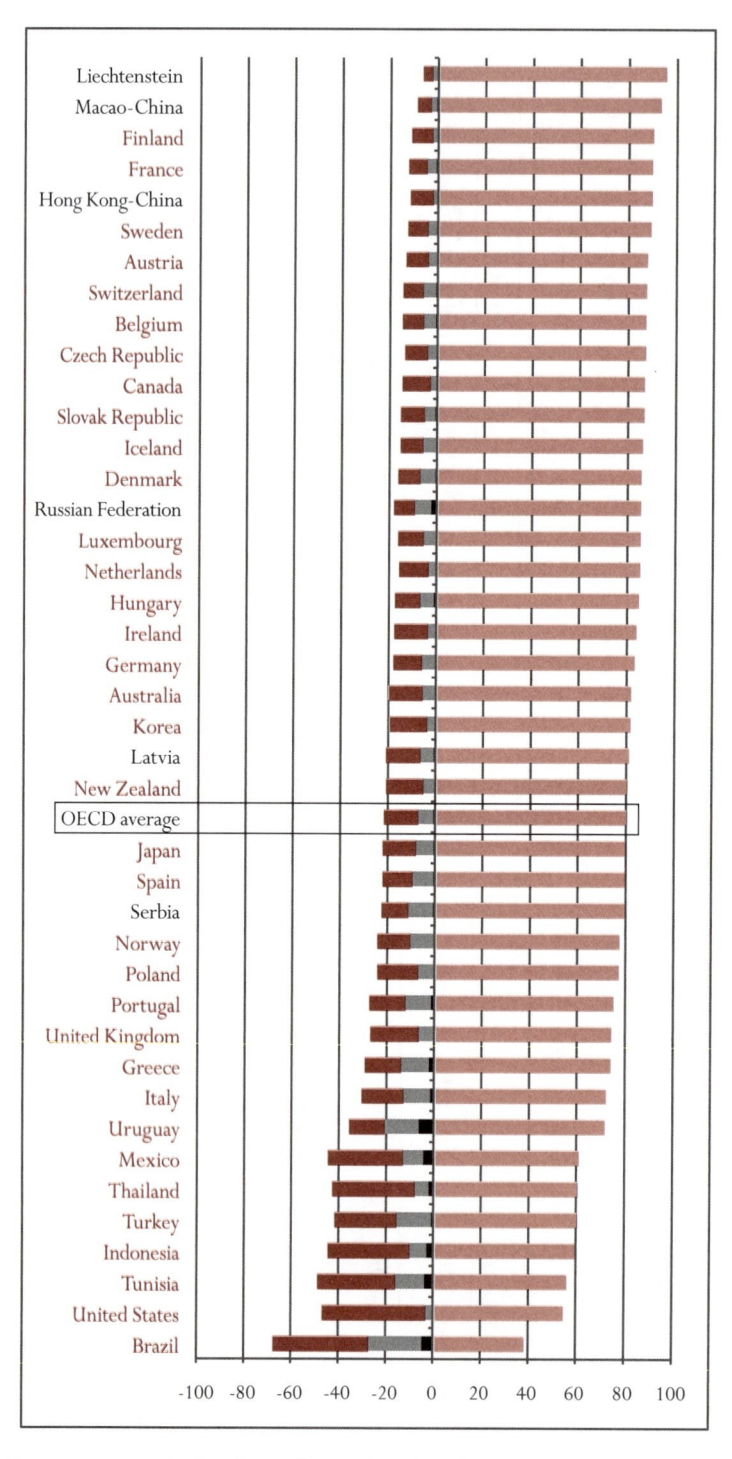

Figure A1.1 ■ **Student performance on Exchange Rate – Question 1**

Percentage of students who:

■ Gave an incorrect answer   ■ Did not reach the question
■ Did not answer the question ■ Answered correctly – 12600 ZAR (unit not required)

Note: Countries are ranked in descending order of students who gave the correct answer.

200 Learning Mathematics for Life: A Perspective from PISA – © OECD 2009

Figure A1.2 ■ **Student performance on Staircase – Question 1**

Percentage of students who:
- Gave an incorrect answer
- Did not answer the question
- Did not reach the question
- Answered correctly – 18

Macao-China
Hong Kong-China
Switzerland
Finland
Liechtenstein
Netherlands
Poland
France
Latvia
Austria
Italy
Belgium
Denmark
Sweden
Norway
Canada
Japan
Korea
Ireland
New Zealand
Czech Republic
Australia
Spain
Iceland
OECD average
Portugal
Slovak Republic
Serbia
Germany
Russian Federation
Luxembourg
United Kingdom
Greece
United States
Uruguay
Mexico
Turkey
Hungary
Tunisia
Thailand
Indonesia
Brazil

Note: Countries are ranked in descending order of students who gave the correct answer.

Learning Mathematics for Life: A Perspective from PISA – © OECD 2009

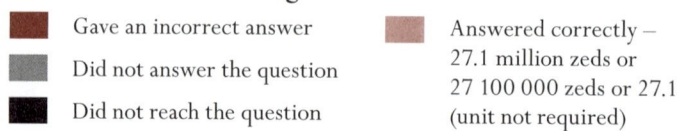

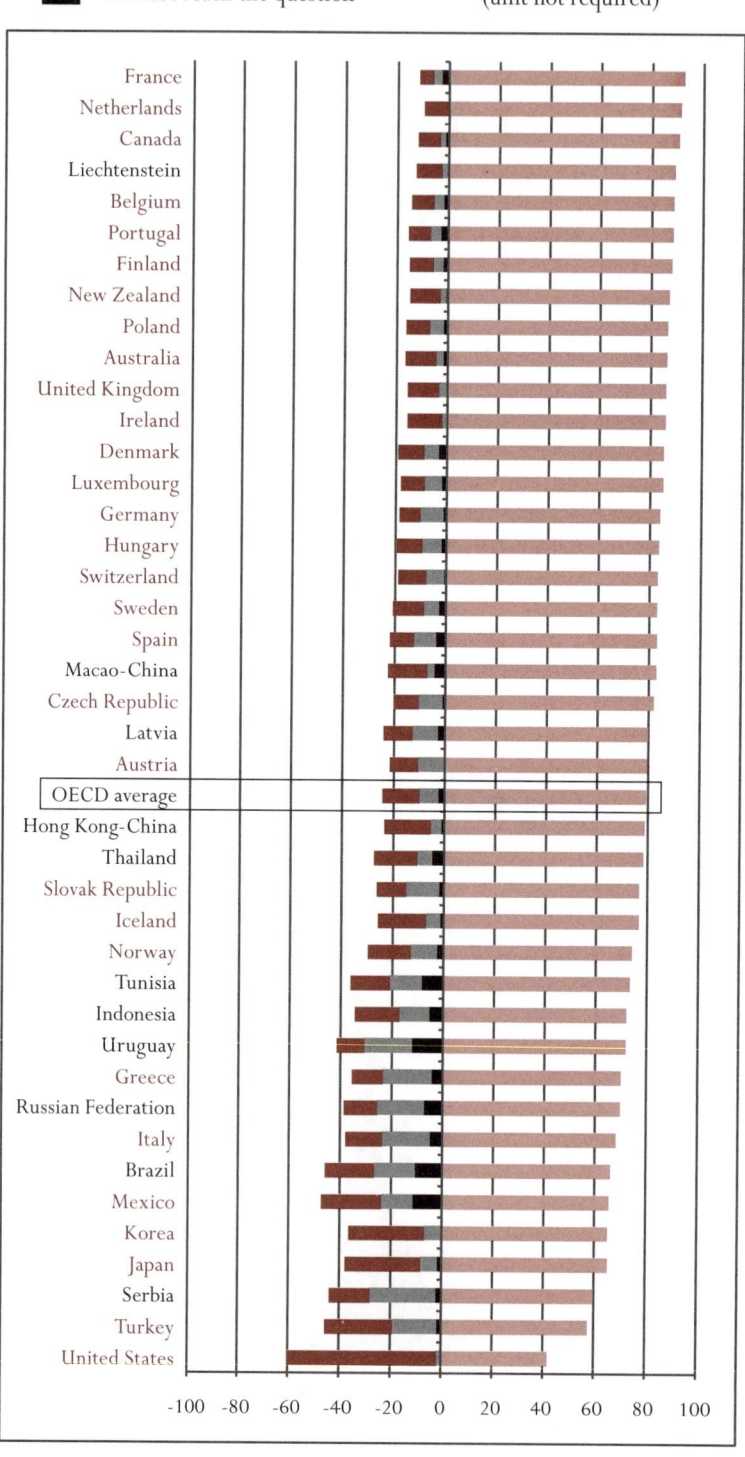

Figure A1.3 ■ Student performance on Exports – Question 1

Percentage of students who:

- Gave an incorrect answer
- Did not answer the question
- Did not reach the question
- Answered correctly – 27.1 million zeds or 27 100 000 zeds or 27.1 (unit not required)

France
Netherlands
Canada
Liechtenstein
Belgium
Portugal
Finland
New Zealand
Poland
Australia
United Kingdom
Ireland
Denmark
Luxembourg
Germany
Hungary
Switzerland
Sweden
Spain
Macao-China
Czech Republic
Latvia
Austria
OECD average
Hong Kong-China
Thailand
Slovak Republic
Iceland
Norway
Tunisia
Indonesia
Uruguay
Greece
Russian Federation
Italy
Brazil
Mexico
Korea
Japan
Serbia
Turkey
United States

-100 -80 -60 -40 -20 0 20 40 60 80 100

Note: Countries are ranked in descending order of students who gave the correct answer.

Figure A1.4 ■ **Student performance on Exchange Rate – Question 2**

Percentage of students who:
- Gave an incorrect answer
- Did not answer the question
- Did not reach the question
- Answered correctly – 975 SGD (unit not required)

Note: Countries are ranked in descending order of students who gave the correct answer.

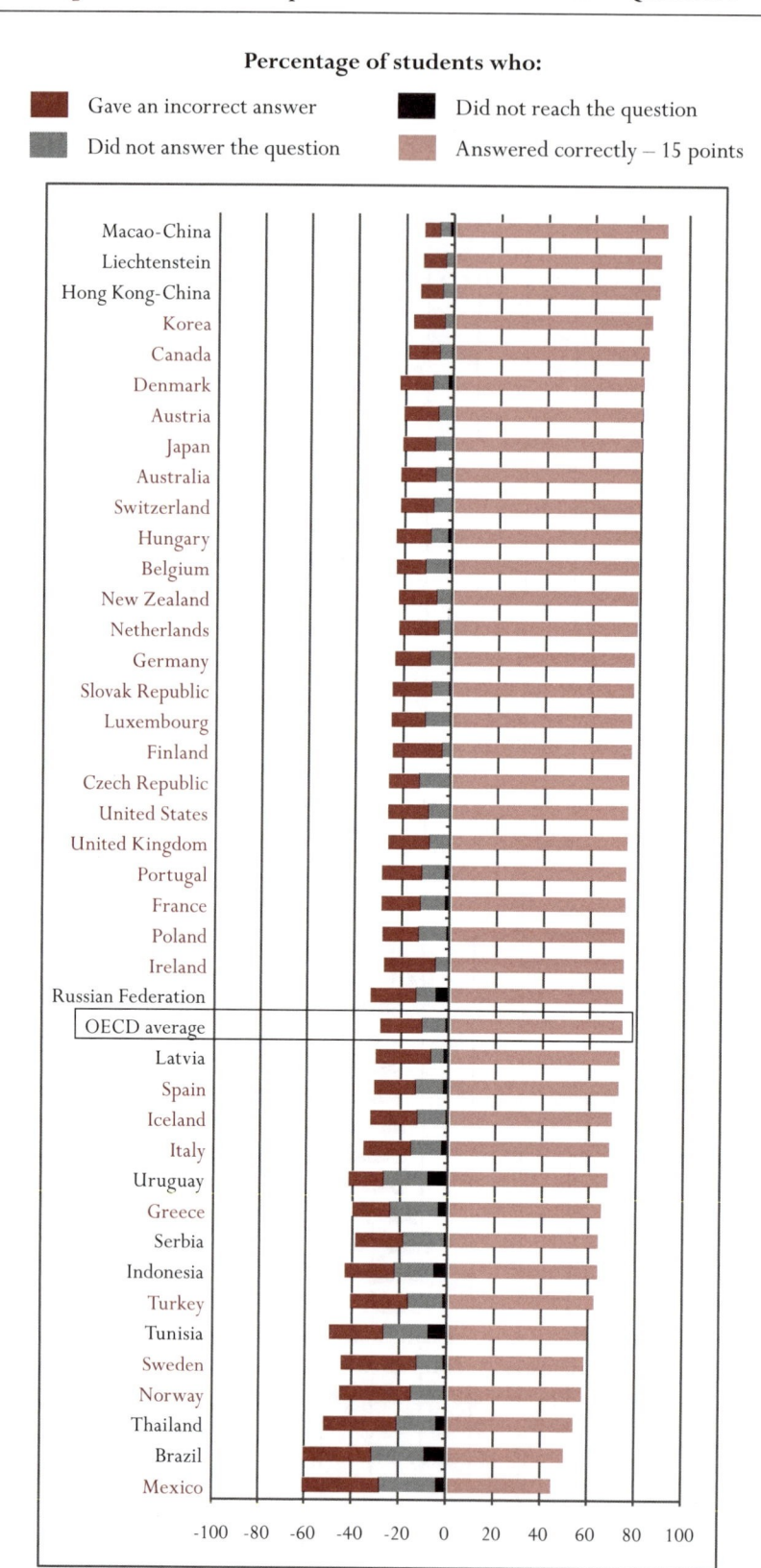

Figure A1.5 ■ **Student performance on The Best Car – Question 1**

Percentage of students who:

- Gave an incorrect answer
- Did not reach the question
- Did not answer the question
- Answered correctly – 15 points

Macao-China
Liechtenstein
Hong Kong-China
Korea
Canada
Denmark
Austria
Japan
Australia
Switzerland
Hungary
Belgium
New Zealand
Netherlands
Germany
Slovak Republic
Luxembourg
Finland
Czech Republic
United States
United Kingdom
Portugal
France
Poland
Ireland
Russian Federation
OECD average
Latvia
Spain
Iceland
Italy
Uruguay
Greece
Serbia
Indonesia
Turkey
Tunisia
Sweden
Norway
Thailand
Brazil
Mexico

-100 -80 -60 -40 -20 0 20 40 60 80 100

Note: Countries are ranked in descending order of students who gave the correct answer.

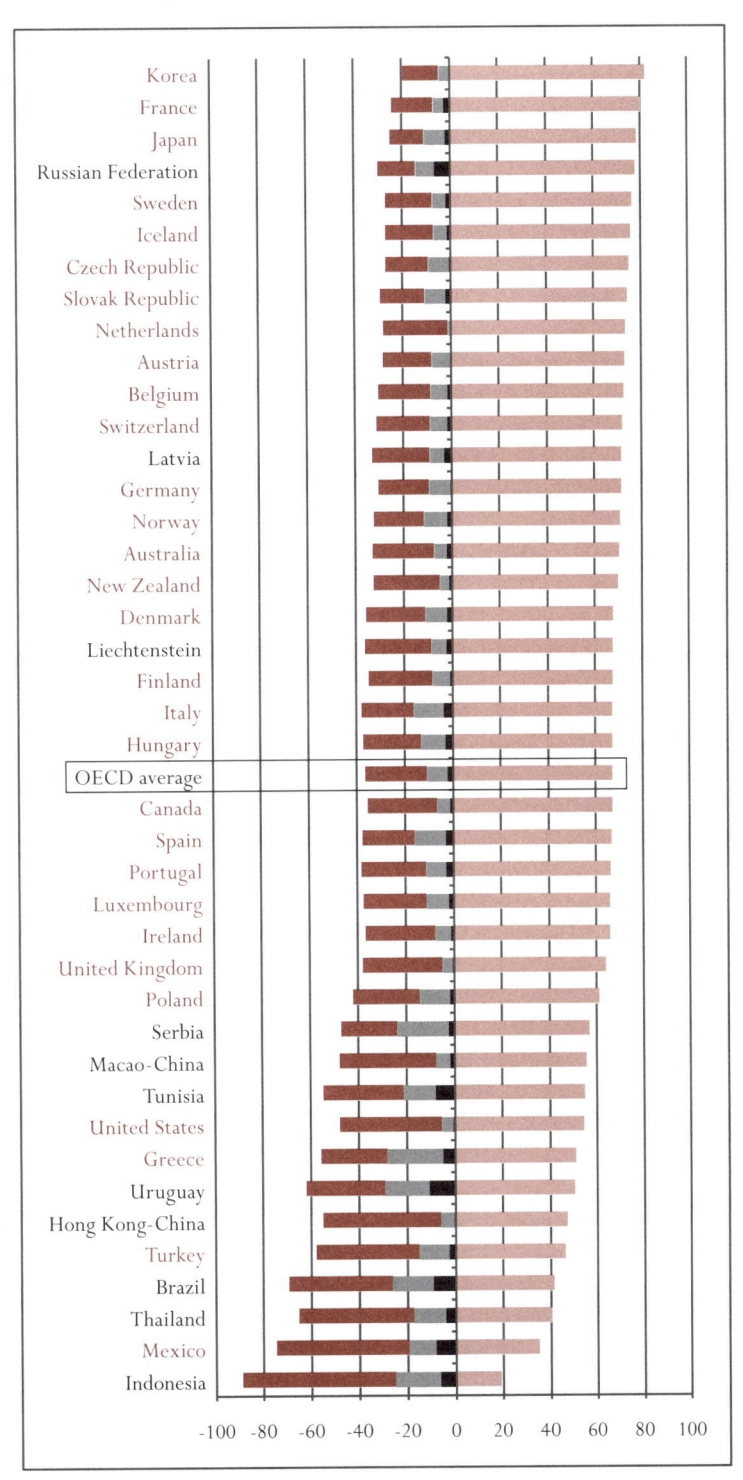

Figure A1.6 ■ **Student performance on Growing Up – Question 1**

Percentage of students who:

- Gave an incorrect answer
- Did not reach the question
- Did not answer the question
- Answered correctly – 168.3 cm (unit already given)

Korea
France
Japan
Russian Federation
Sweden
Iceland
Czech Republic
Slovak Republic
Netherlands
Austria
Belgium
Switzerland
Latvia
Germany
Norway
Australia
New Zealand
Denmark
Liechtenstein
Finland
Italy
Hungary
OECD average
Canada
Spain
Portugal
Luxembourg
Ireland
United Kingdom
Poland
Serbia
Macao-China
Tunisia
United States
Greece
Uruguay
Hong Kong-China
Turkey
Brazil
Thailand
Mexico
Indonesia

-100 -80 -60 -40 -20 0 20 40 60 80 100

Note: Countries are ranked in descending order of students who gave the correct answer.

Learning Mathematics for Life: A Perspective from PISA – © OECD 2009

Figure A1.7 ■ **Student performance on Growing Up – Question 2**

Percentage of students who:

- Gave an incorrect answer (1998, Girls are taller than boys when they're older than 13 years)
- Did not answer the question
- Did not reach the question
- Answered correctly – gave the correct interval (from 11 to 13 years) or stated that girls are taller than boys when they are 11 and 12 years old
- Gave answer of a subset of 11, 12 and 13 (12 to 13, 12, 13, 11, 11.2 to 12.8)

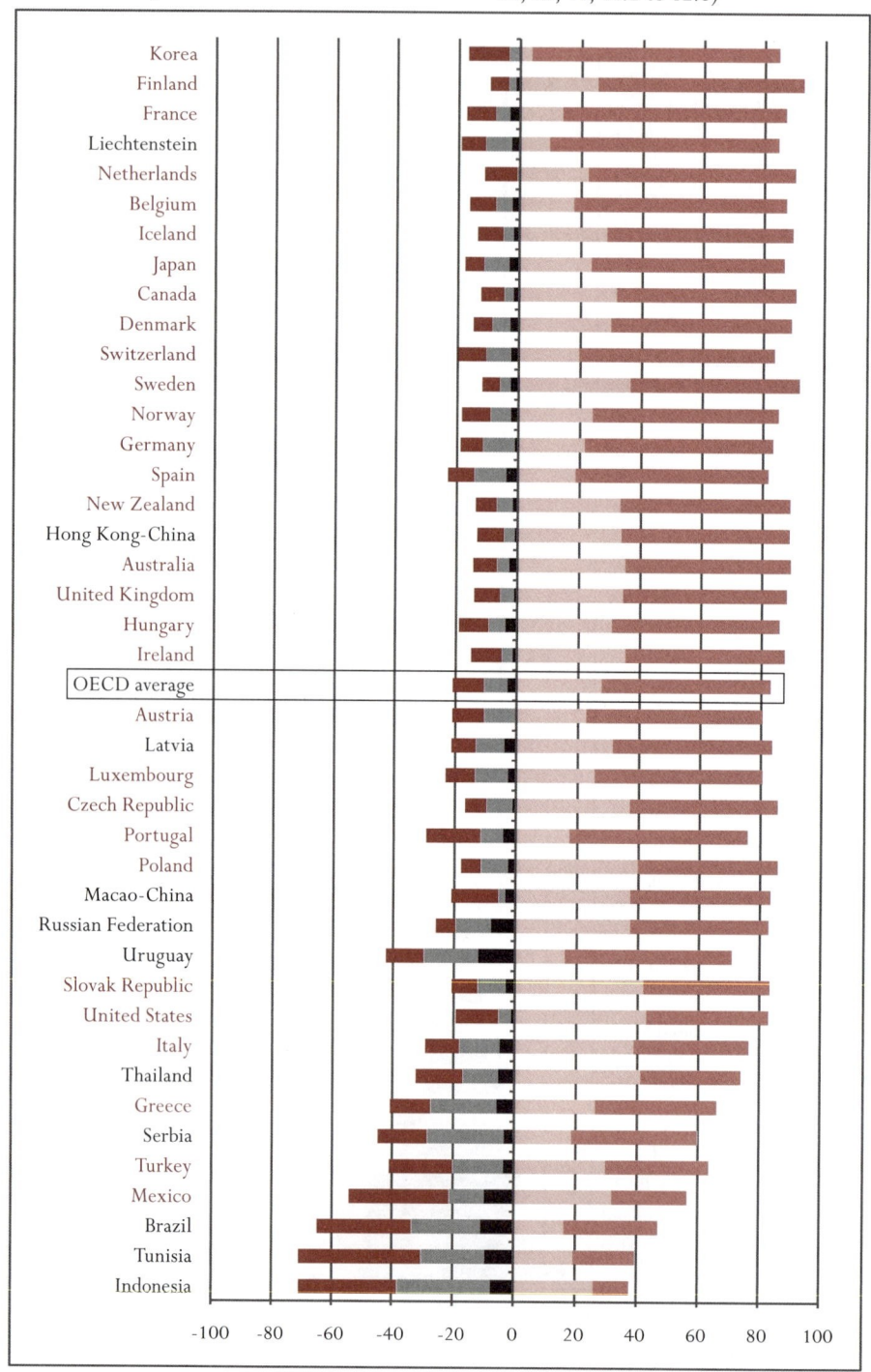

Countries (top to bottom): Korea, Finland, France, Liechtenstein, Netherlands, Belgium, Iceland, Japan, Canada, Denmark, Switzerland, Sweden, Norway, Germany, Spain, New Zealand, Hong Kong-China, Australia, United Kingdom, Hungary, Ireland, OECD average, Austria, Latvia, Luxembourg, Czech Republic, Portugal, Poland, Macao-China, Russian Federation, Uruguay, Slovak Republic, United States, Italy, Thailand, Greece, Serbia, Turkey, Mexico, Brazil, Tunisia, Indonesia

Note: Countries are ranked in descending order of total percent correct, that is, allowing for both full and partial credits.

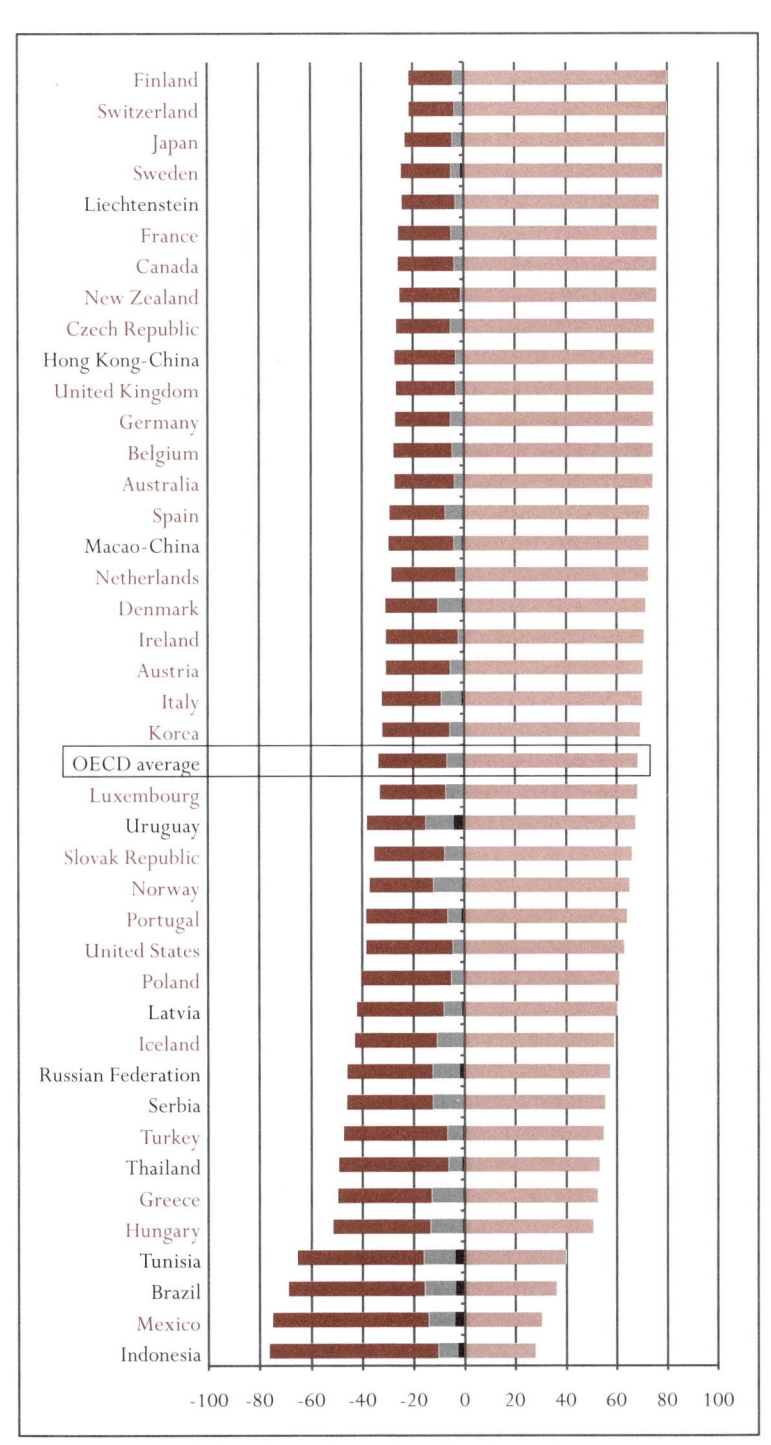

Figure A1.8 ■ **Student performance on Cubes – Question 1**

Percentage of students who:

- Gave an incorrect answer
- Did not answer the question
- Did not reach the question
- Answered correctly – Top row (1 5 4), Bottom row (2 6 5). Equivalent answer shown as dice faces is also acceptable.

Finland
Switzerland
Japan
Sweden
Liechtenstein
France
Canada
New Zealand
Czech Republic
Hong Kong-China
United Kingdom
Germany
Belgium
Australia
Spain
Macao-China
Netherlands
Denmark
Ireland
Austria
Italy
Korea
OECD average
Luxembourg
Uruguay
Slovak Republic
Norway
Portugal
United States
Poland
Latvia
Iceland
Russian Federation
Serbia
Turkey
Thailand
Greece
Hungary
Tunisia
Brazil
Mexico
Indonesia

Note: Countries are ranked in descending order of students who gave the correct answer.

Learning Mathematics for Life: A Perspective from PISA – © OECD 2009

207

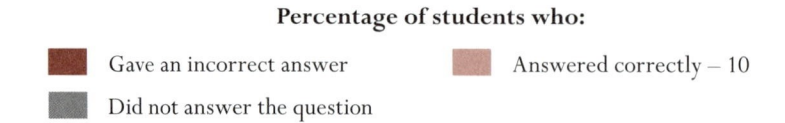

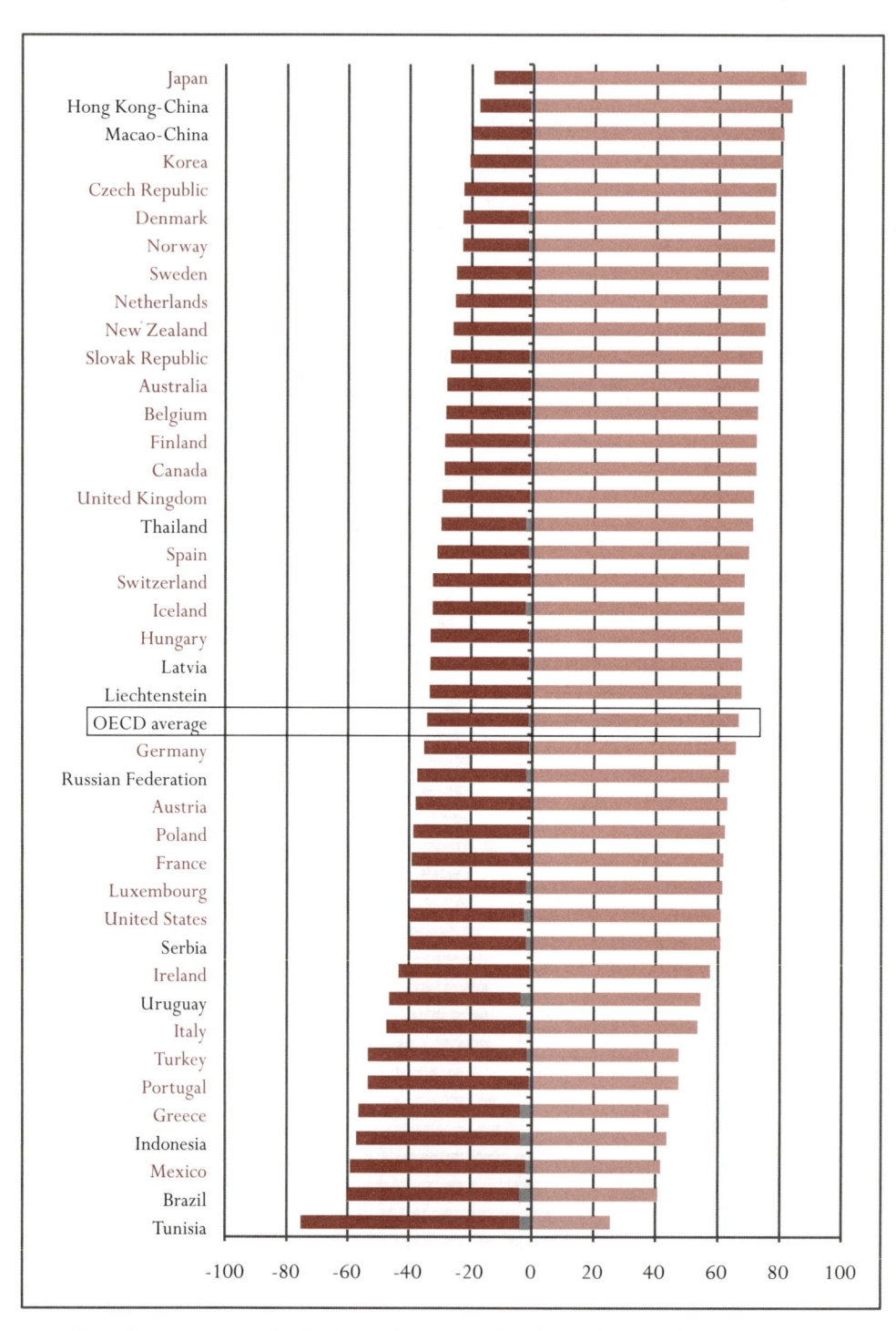

Figure A1.9 ■ **Student performance on Step Pattern – Question 1**

Percentage of students who:
- Gave an incorrect answer
- Answered correctly – 10
- Did not answer the question

Note: Countries are ranked in descending order of students who gave the correct answer.

Figure A1.10 ■ **Student performance on Skateboard – Question 1**

Percentage of students who:
- Did not get either the minimum or maximum price correct
- Did not answer the question
- Did not reach the question
- Got both the minimum and maximum correct
- Got either only the minimum or only the maximum correct

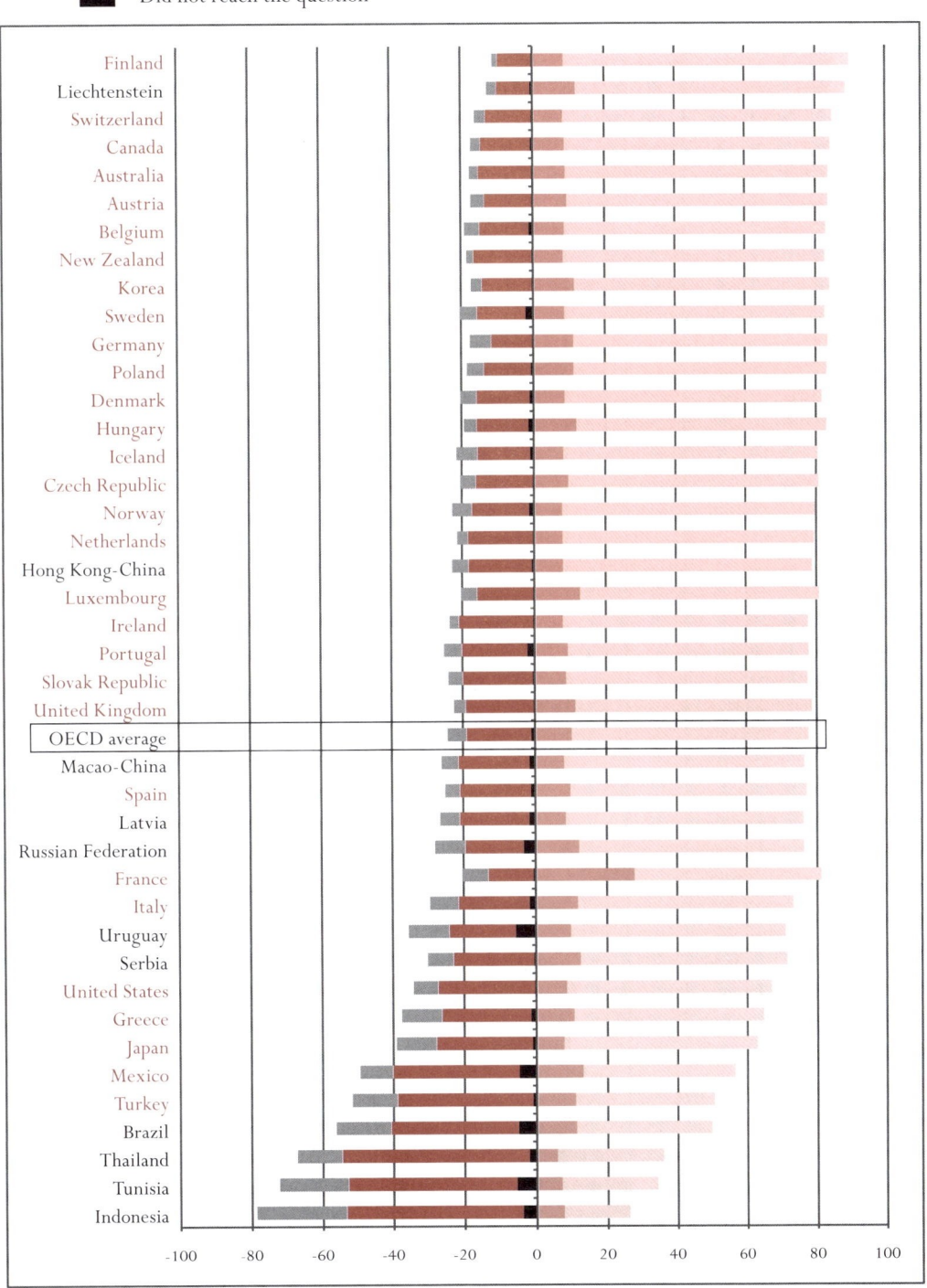

Note: Countries are ranked in descending order of total percent correct, that is, percent full credit + half of the percent partial credit.

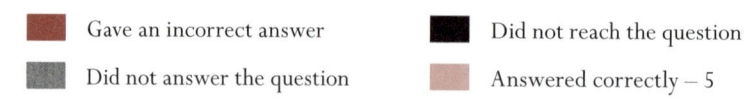

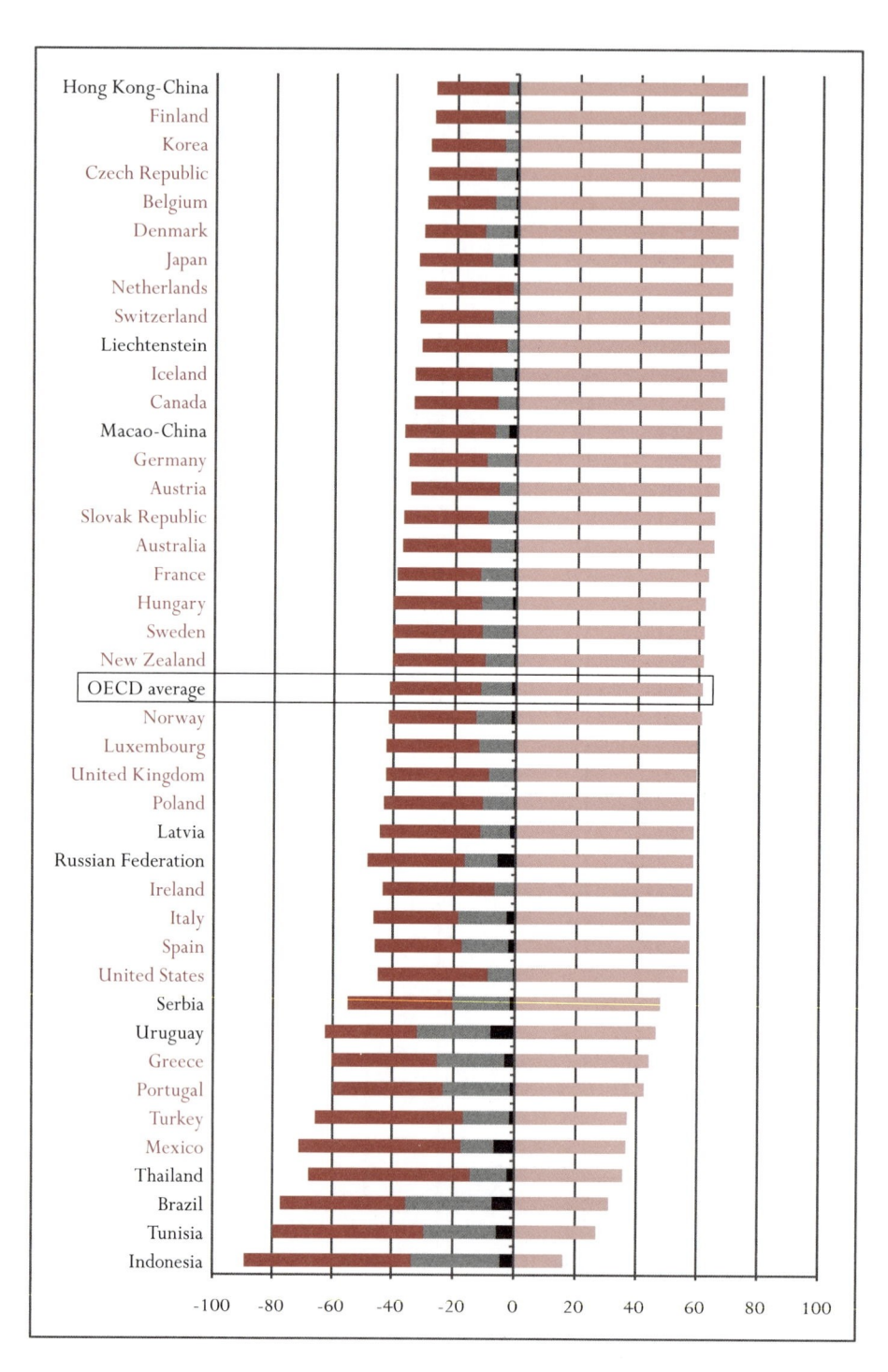

Figure A1.11 ■ **Student performance on Bookshelves – Question 1**

Percentage of students who:

- Gave an incorrect answer
- Did not reach the question
- Did not answer the question
- Answered correctly – 5

Hong Kong-China
Finland
Korea
Czech Republic
Belgium
Denmark
Japan
Netherlands
Switzerland
Liechtenstein
Iceland
Canada
Macao-China
Germany
Austria
Slovak Republic
Australia
France
Hungary
Sweden
New Zealand
OECD average
Norway
Luxembourg
United Kingdom
Poland
Latvia
Russian Federation
Ireland
Italy
Spain
United States
Serbia
Uruguay
Greece
Portugal
Turkey
Mexico
Thailand
Brazil
Tunisia
Indonesia

-100 -80 -60 -40 -20 0 20 40 60 80 100

Note: Countries are ranked in descending order of students who gave the correct answer.

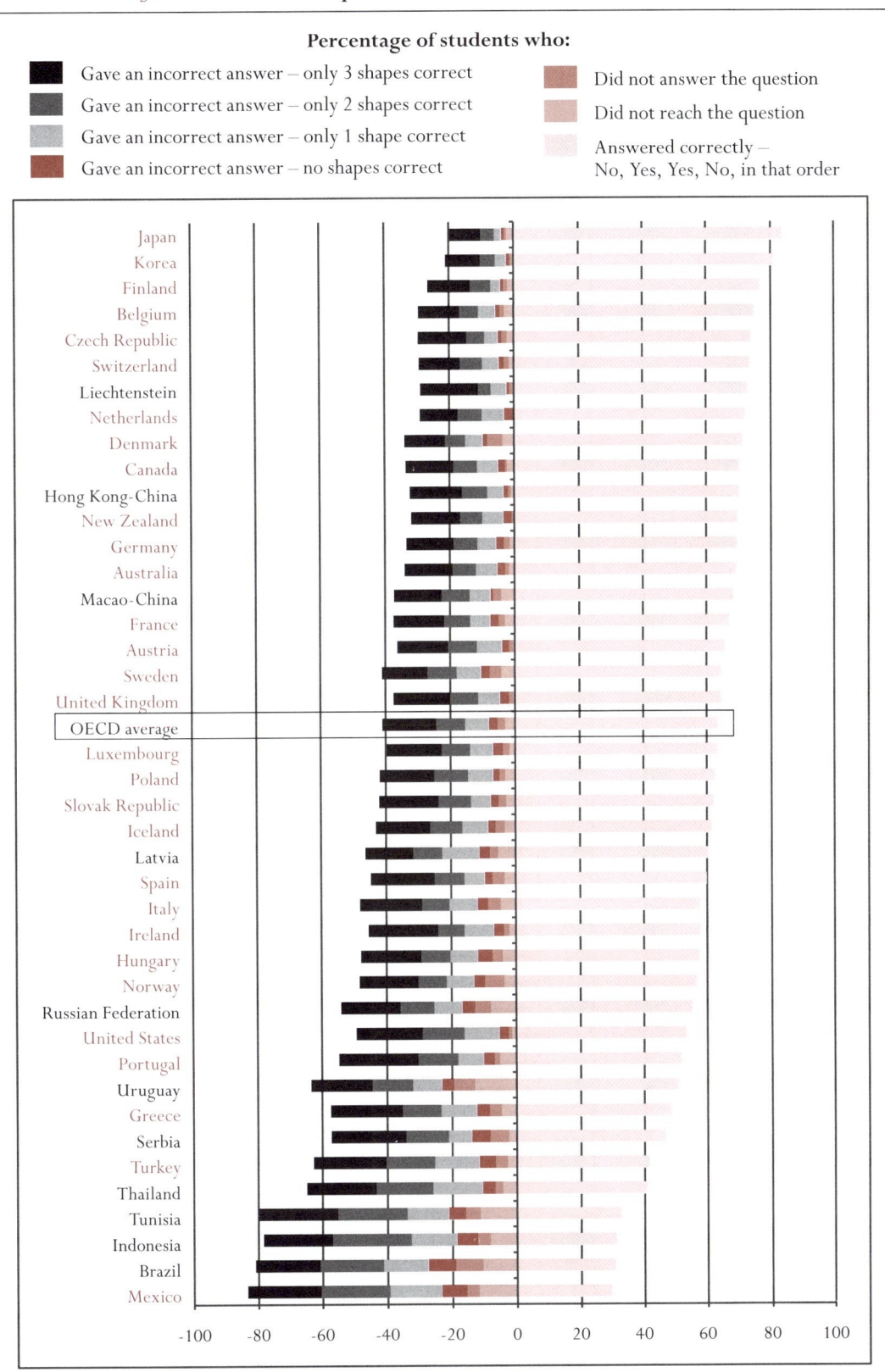

Figure A1.12 ■ **Student performance on Number Cubes – Question 2**

Percentage of students who:
- Gave an incorrect answer – only 3 shapes correct
- Gave an incorrect answer – only 2 shapes correct
- Gave an incorrect answer – only 1 shape correct
- Gave an incorrect answer – no shapes correct
- Did not answer the question
- Did not reach the question
- Answered correctly – No, Yes, Yes, No, in that order

Note: Countries are ranked in descending order of students who gave the correct answer.

Learning Mathematics for Life: A Perspective from PISA – © OECD 2009

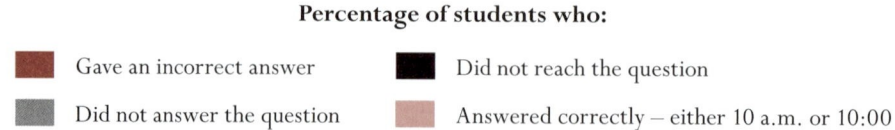

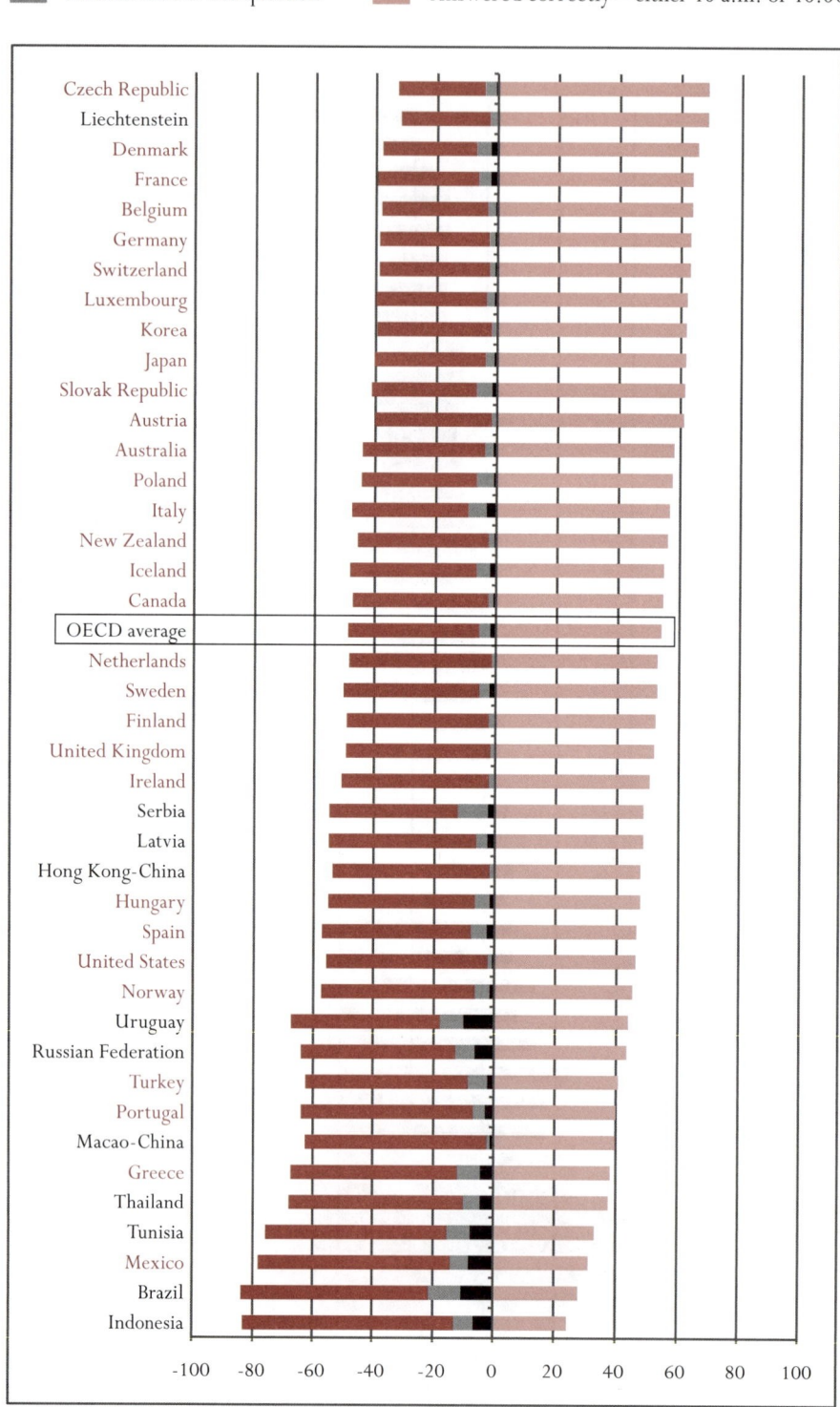

Figure A1.13 ■ **Student performance on Internet Relay Chat – Question 1**

Percentage of students who:

- Gave an incorrect answer
- Did not reach the question
- Did not answer the question
- Answered correctly – either 10 a.m. or 10:00

Note: Countries are ranked in descending order of students who gave the correct answer.

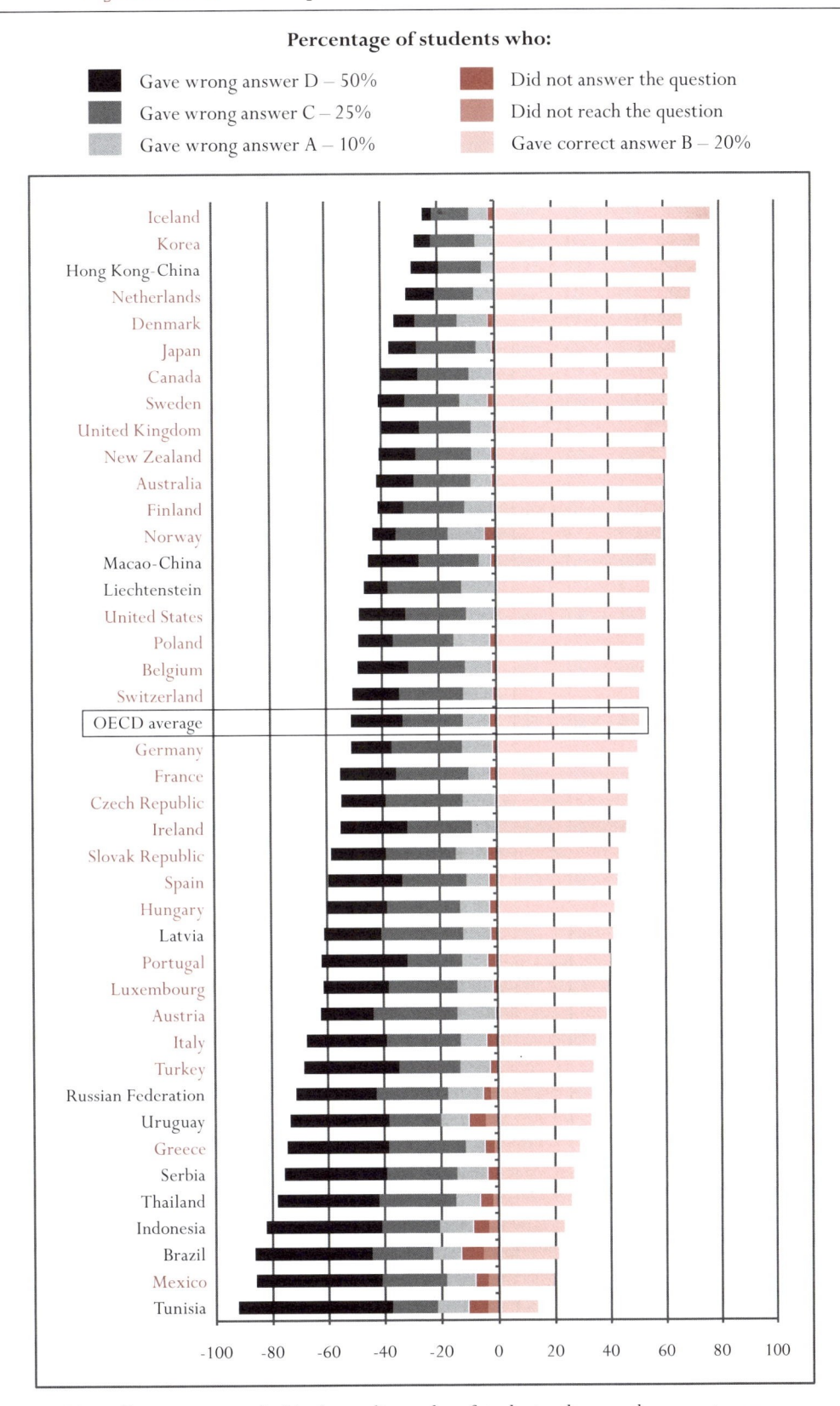

Figure A1.14 ■ **Student performance on Coloured Candies – Question 1**

Percentage of students who:

- Gave wrong answer D – 50%
- Gave wrong answer C – 25%
- Gave wrong answer A – 10%
- Did not answer the question
- Did not reach the question
- Gave correct answer B – 20%

Iceland
Korea
Hong Kong-China
Netherlands
Denmark
Japan
Canada
Sweden
United Kingdom
New Zealand
Australia
Finland
Norway
Macao-China
Liechtenstein
United States
Poland
Belgium
Switzerland
OECD average
Germany
France
Czech Republic
Ireland
Slovak Republic
Spain
Hungary
Latvia
Portugal
Luxembourg
Austria
Italy
Turkey
Russian Federation
Uruguay
Greece
Serbia
Thailand
Indonesia
Brazil
Mexico
Tunisia

Note: Countries are ranked in descending order of students who gave the correct answer.

Learning Mathematics for Life: A Perspective from PISA – © OECD 2009   213

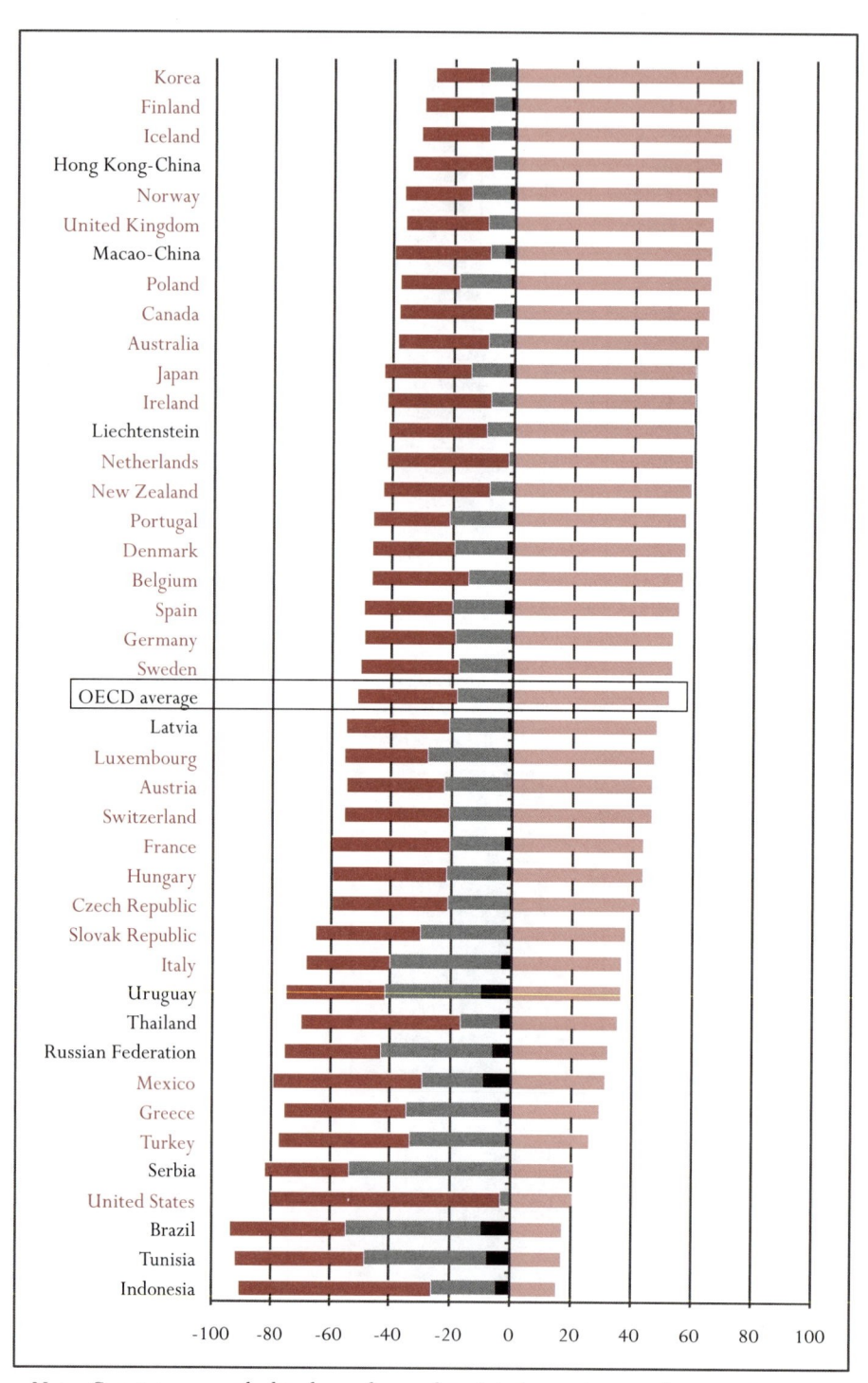

Figure A1.15 ■ **Student performance on Litter – Question 1**

Percentage of students who:
- Gave an incorrect answer
- Did not answer the question
- Did not reach the question
- Answered correctly – reason focuses on big variance in data or on the variability of the data for some categories

Note: Countries are ranked in descending order of students who gave the correct answer.

Figure A1.16 ■ **Student performance on Skateboard – Question 3**

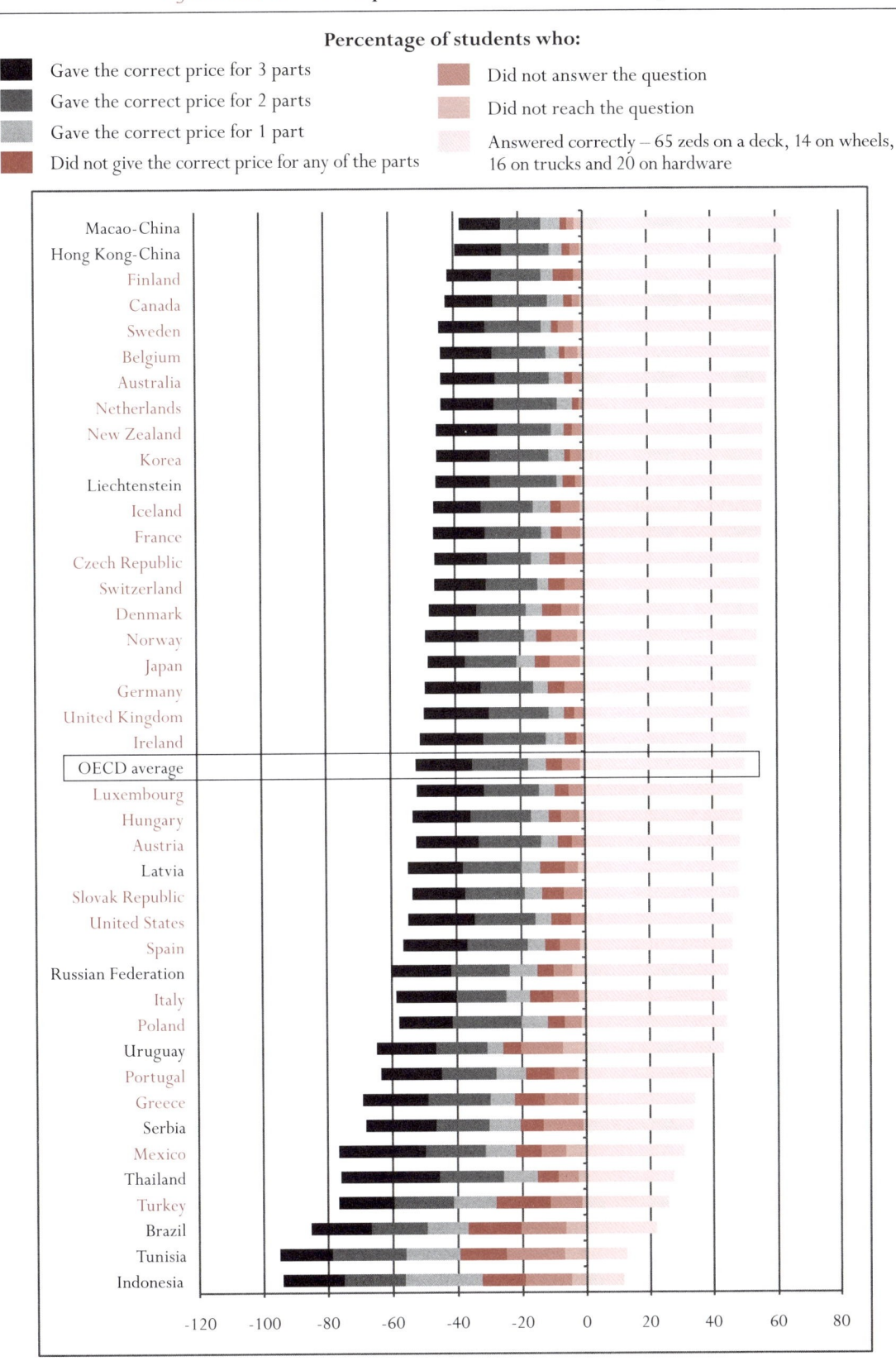

Figure A1.17 ■ **Student performance on Science Tests – Question 1**

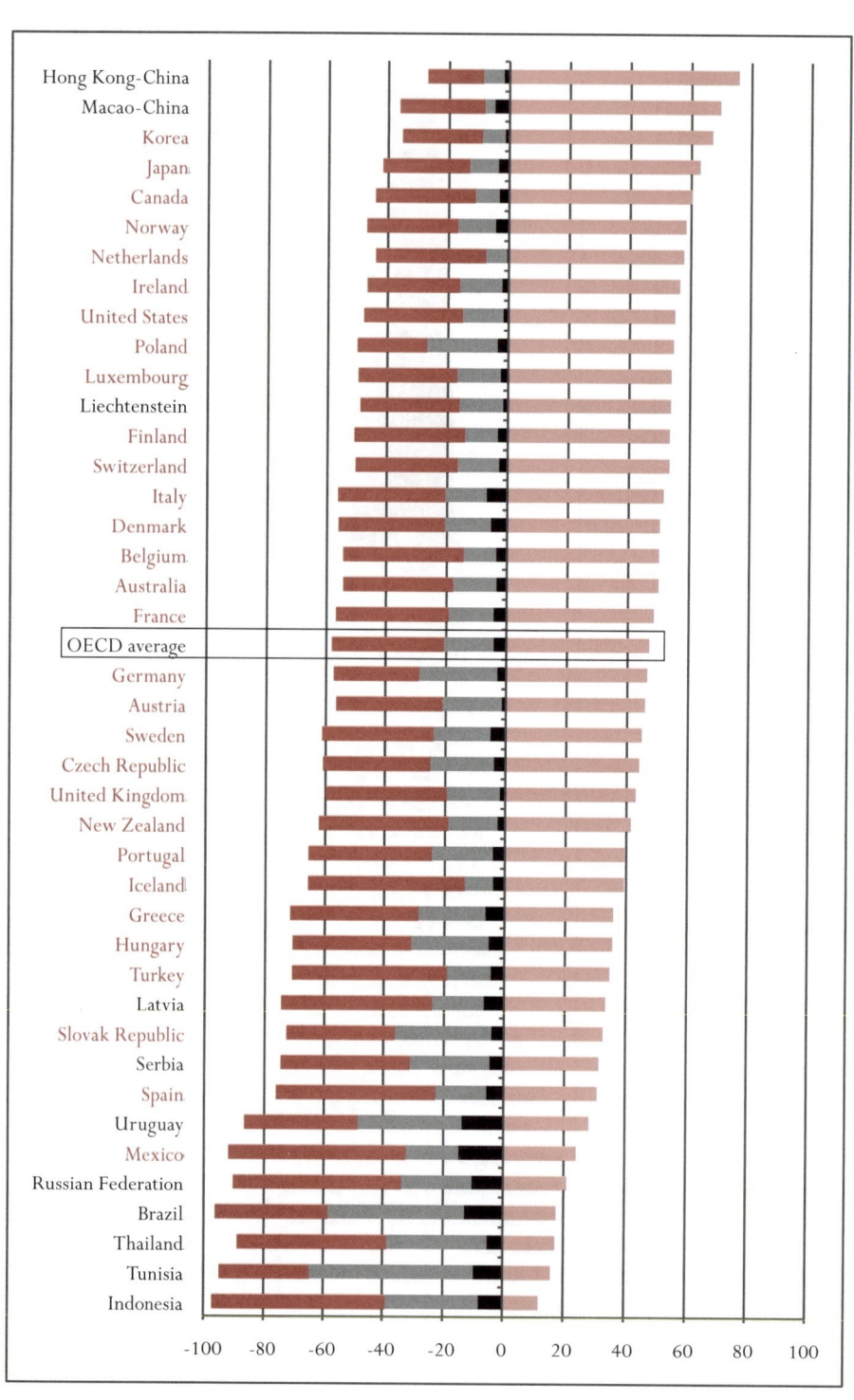

Percentage of students who:

- Gave an incorrect answer
- Did not answer the question
- Did not reach the question
- Answered correctly – 64

Note: Countries are ranked in descending order of students who gave the correct answer.

Annex A1

216 Learning Mathematics for Life: A Perspective from PISA – © OECD 2009

Figure A1.18 ■ **Earthquake – Question 1**

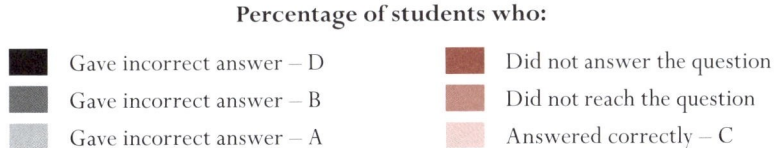

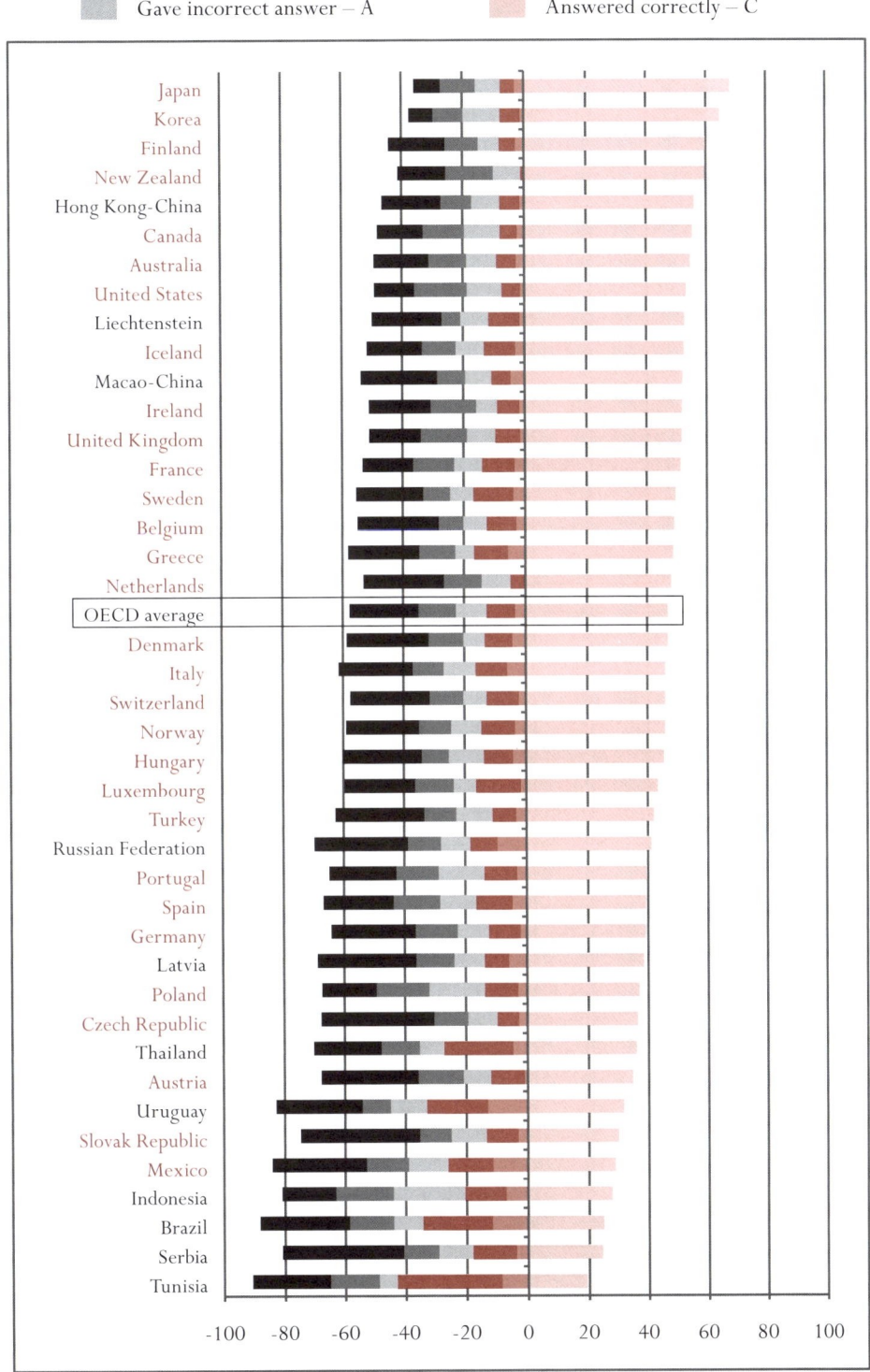

Percentage of students who:

- Gave incorrect answer – D
- Gave incorrect answer – B
- Gave incorrect answer – A
- Did not answer the question
- Did not reach the question
- Answered correctly – C

Note: Countries are ranked in descending order of students who gave the correct answer C.

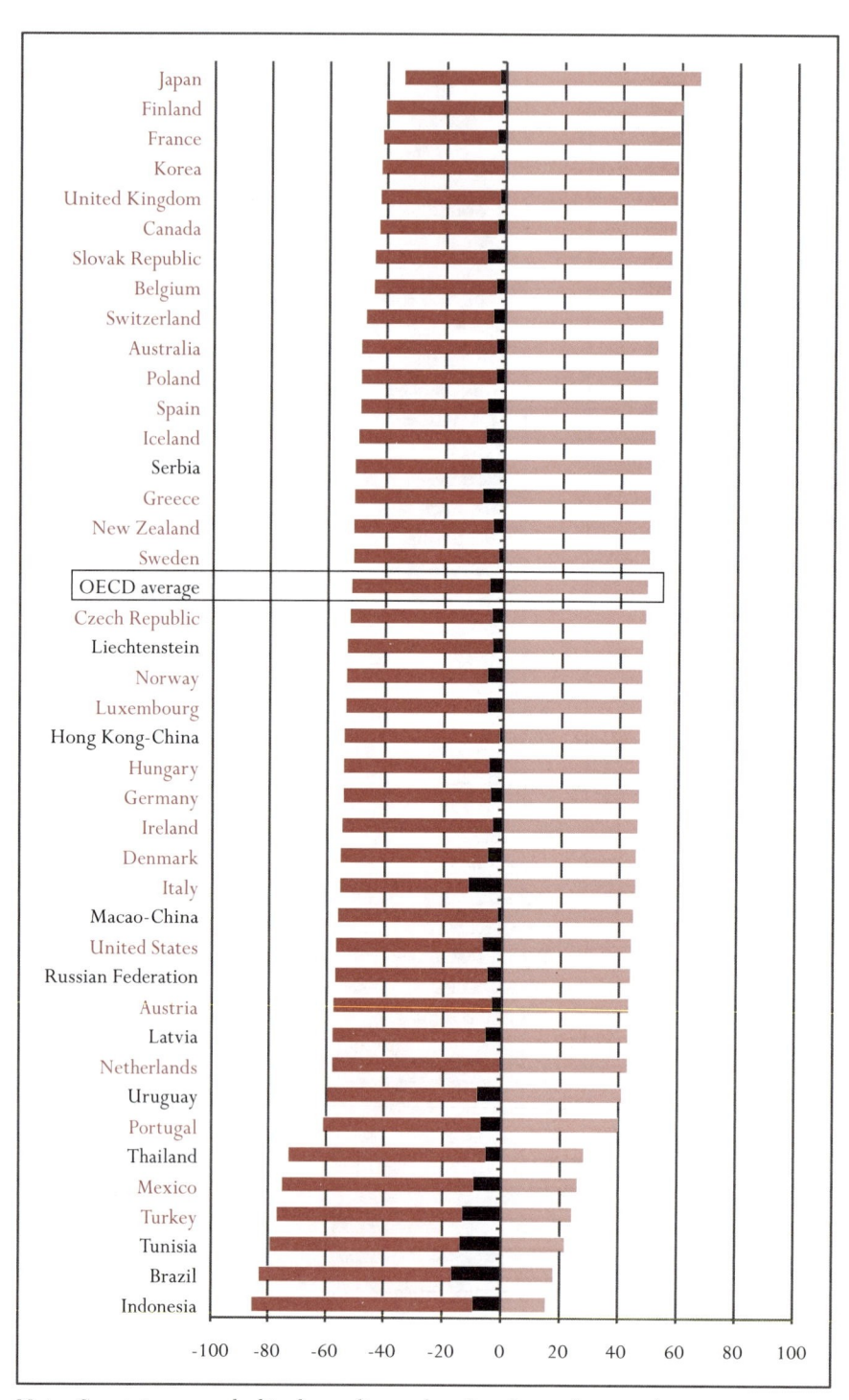

Figure A1.19 ■ **Student performance on Choices – Question 1**

Percentage of students who:
- Gave an incorrect answer
- Answered correctly – 6
- Did not answer the question

Note: Countries are ranked in descending order of students who gave the correct answer.

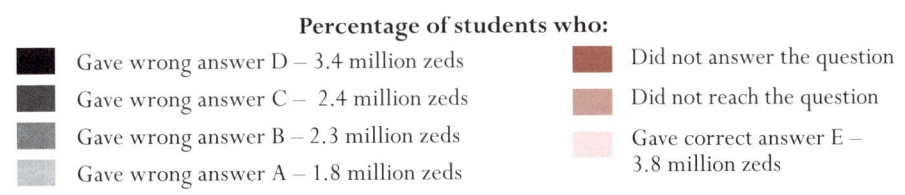

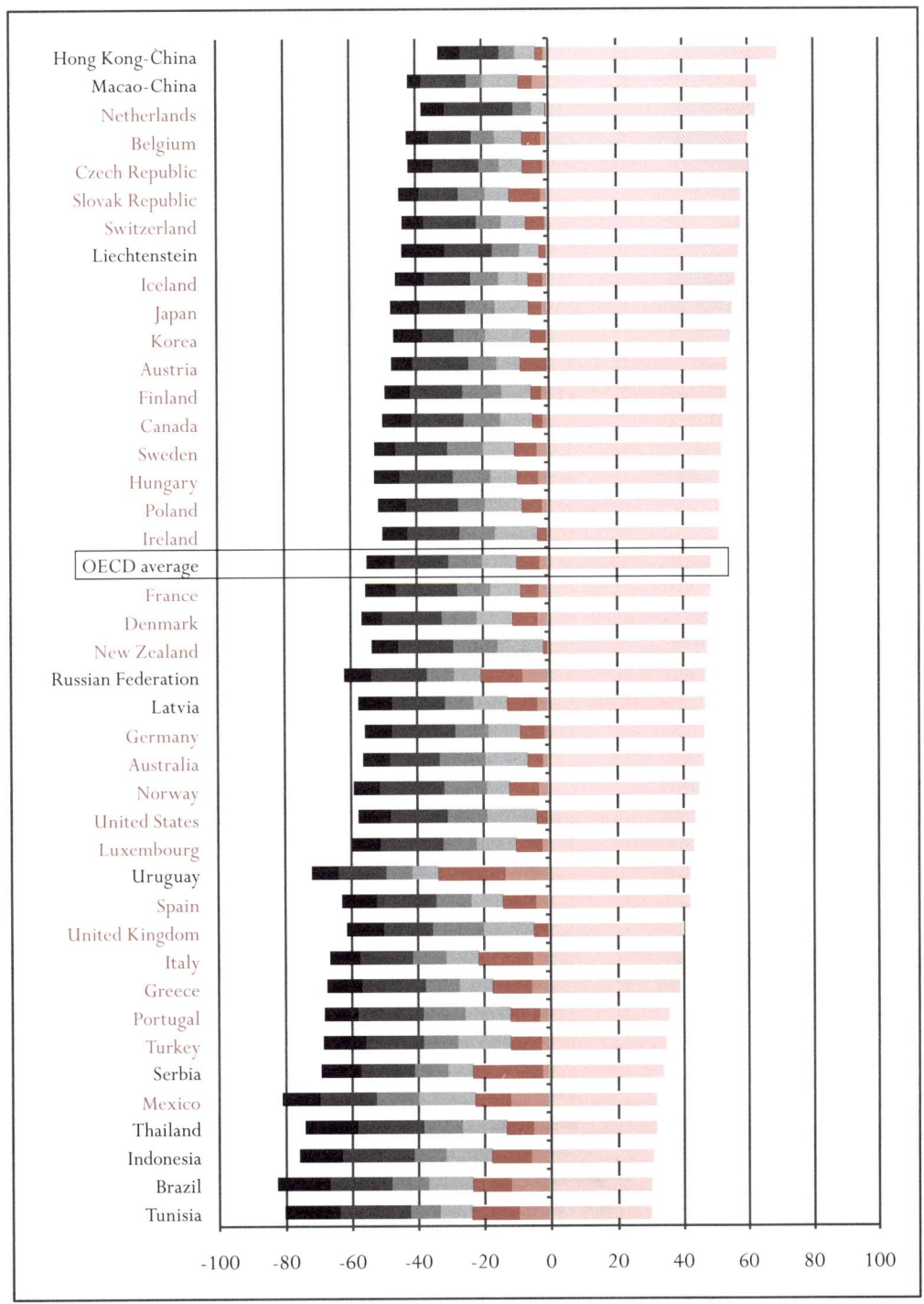

Figure A1.20 ■ **Student performance on Exports – Question 2**

Percentage of students who:

- Gave wrong answer D – 3.4 million zeds
- Gave wrong answer C – 2.4 million zeds
- Gave wrong answer B – 2.3 million zeds
- Gave wrong answer A – 1.8 million zeds
- Did not answer the question
- Did not reach the question
- Gave correct answer E – 3.8 million zeds

Note: Countries are ranked in descending order of students who gave the correct answer E.

Learning Mathematics for Life: A Perspective from PISA – © OECD 2009

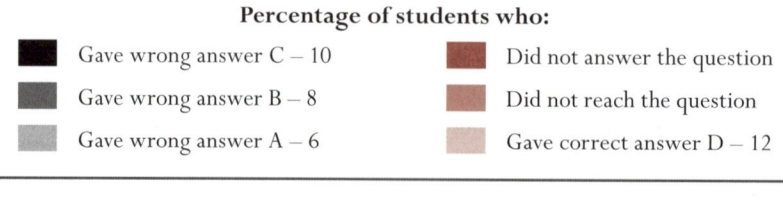

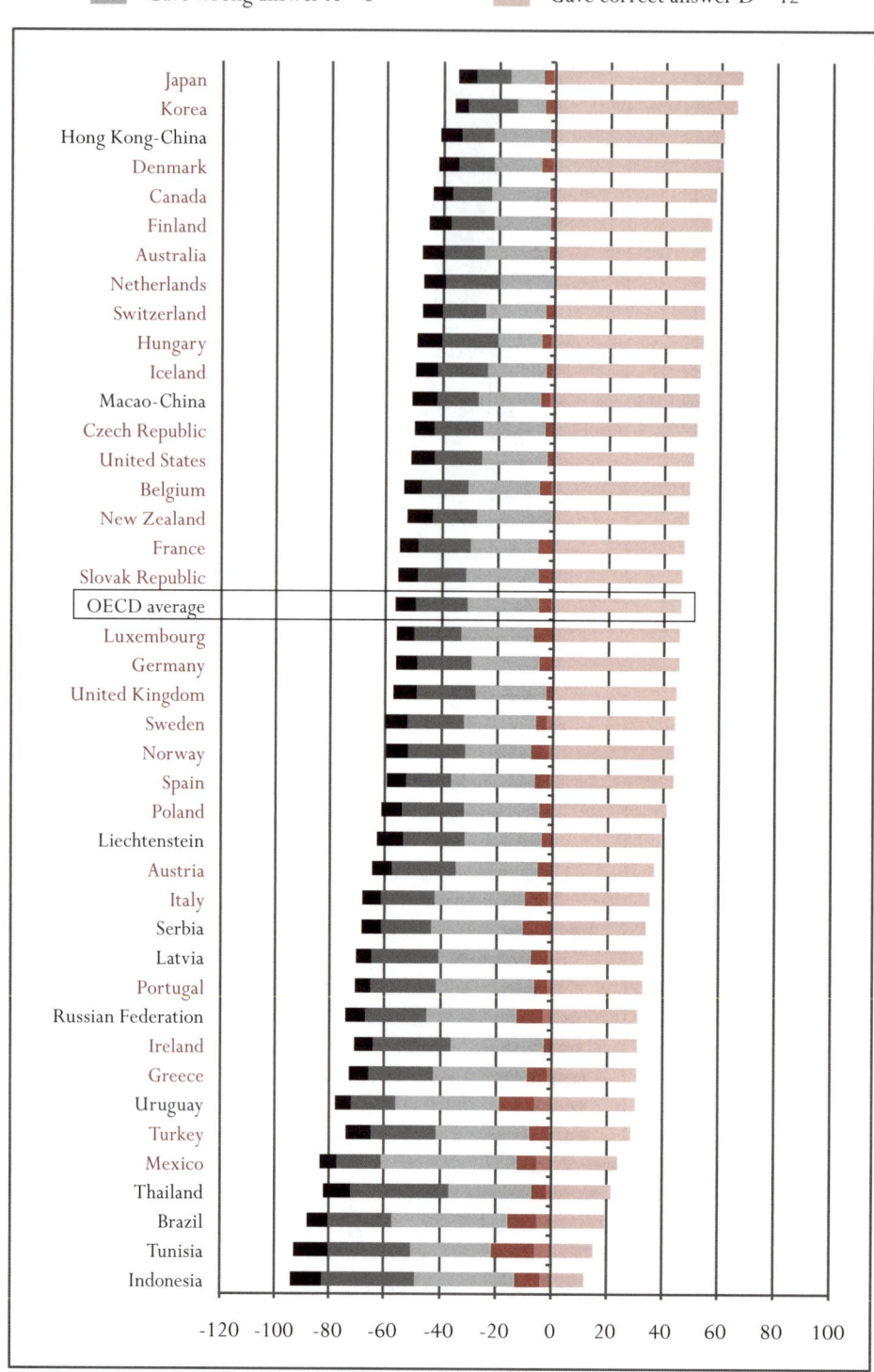

Figure A1.21 ■ **Student performance on Skateboard – Question 2**

Percentage of students who:
- Gave wrong answer C – 10
- Gave wrong answer B – 8
- Gave wrong answer A – 6
- Did not answer the question
- Did not reach the question
- Gave correct answer D – 12

Note: Countries are ranked in descending order of students who gave the correct answer.

Learning Mathematics for Life: A Perspective from PISA – © OECD 2009

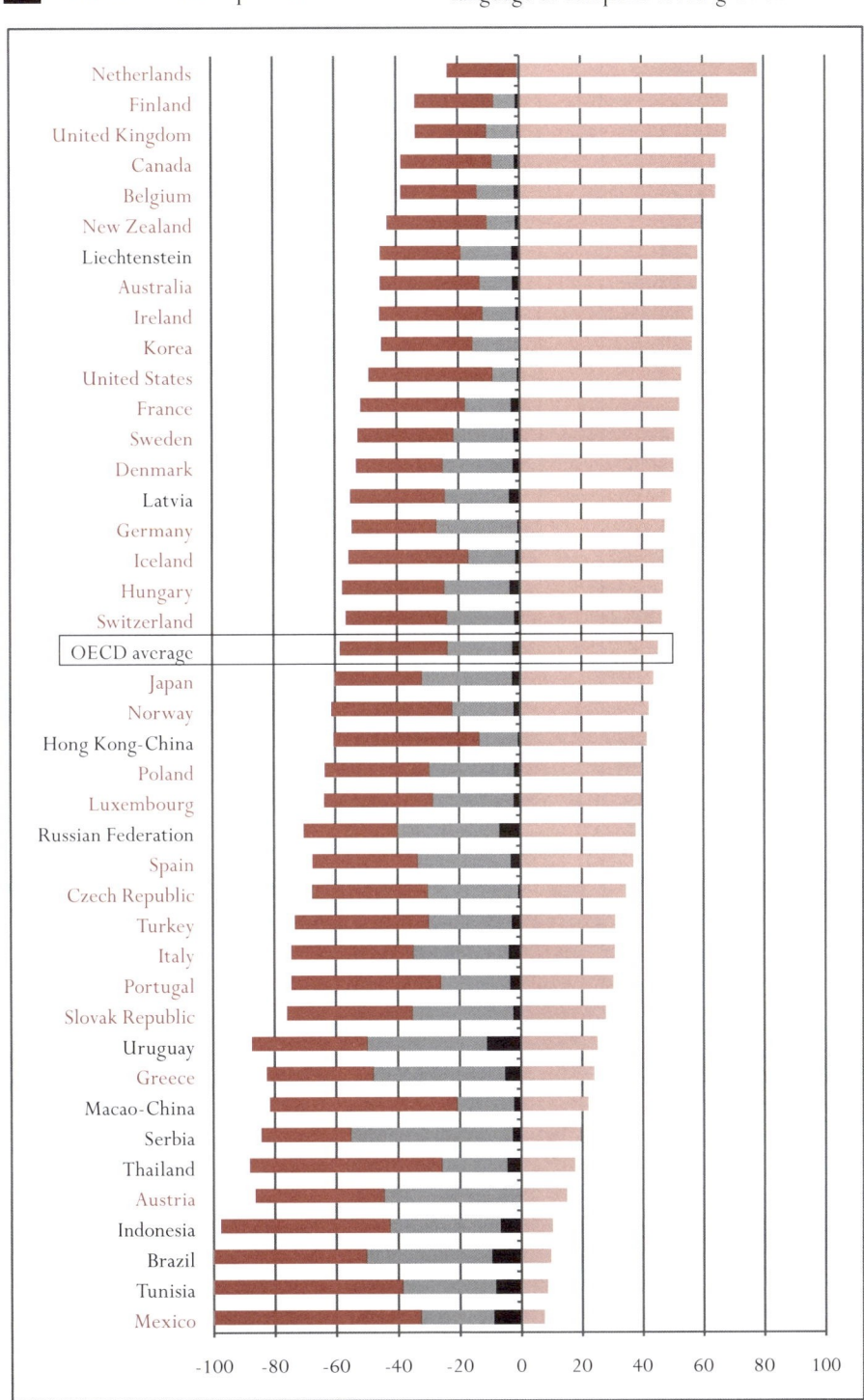

Figure A1.22 ■ **Student performance on Growing Up – Question 3**

Percentage of students who:
- Gave an incorrect answer
- Did not answer the question
- Did not reach the question
- Answered correctly – Refers to the reduced steepness of the curve from 12 years onwards using either daily-life or mathematical language or compares actual growth

Netherlands, Finland, United Kingdom, Canada, Belgium, New Zealand, Liechtenstein, Australia, Ireland, Korea, United States, France, Sweden, Denmark, Latvia, Germany, Iceland, Hungary, Switzerland, OECD average, Japan, Norway, Hong Kong-China, Poland, Luxembourg, Russian Federation, Spain, Czech Republic, Turkey, Italy, Portugal, Slovak Republic, Uruguay, Greece, Macao-China, Serbia, Thailand, Austria, Indonesia, Brazil, Tunisia, Mexico

Note: Countries are ranked in descending order of students who gave the correct answer.

Learning Mathematics for Life: A Perspective from PISA – © OECD 2009

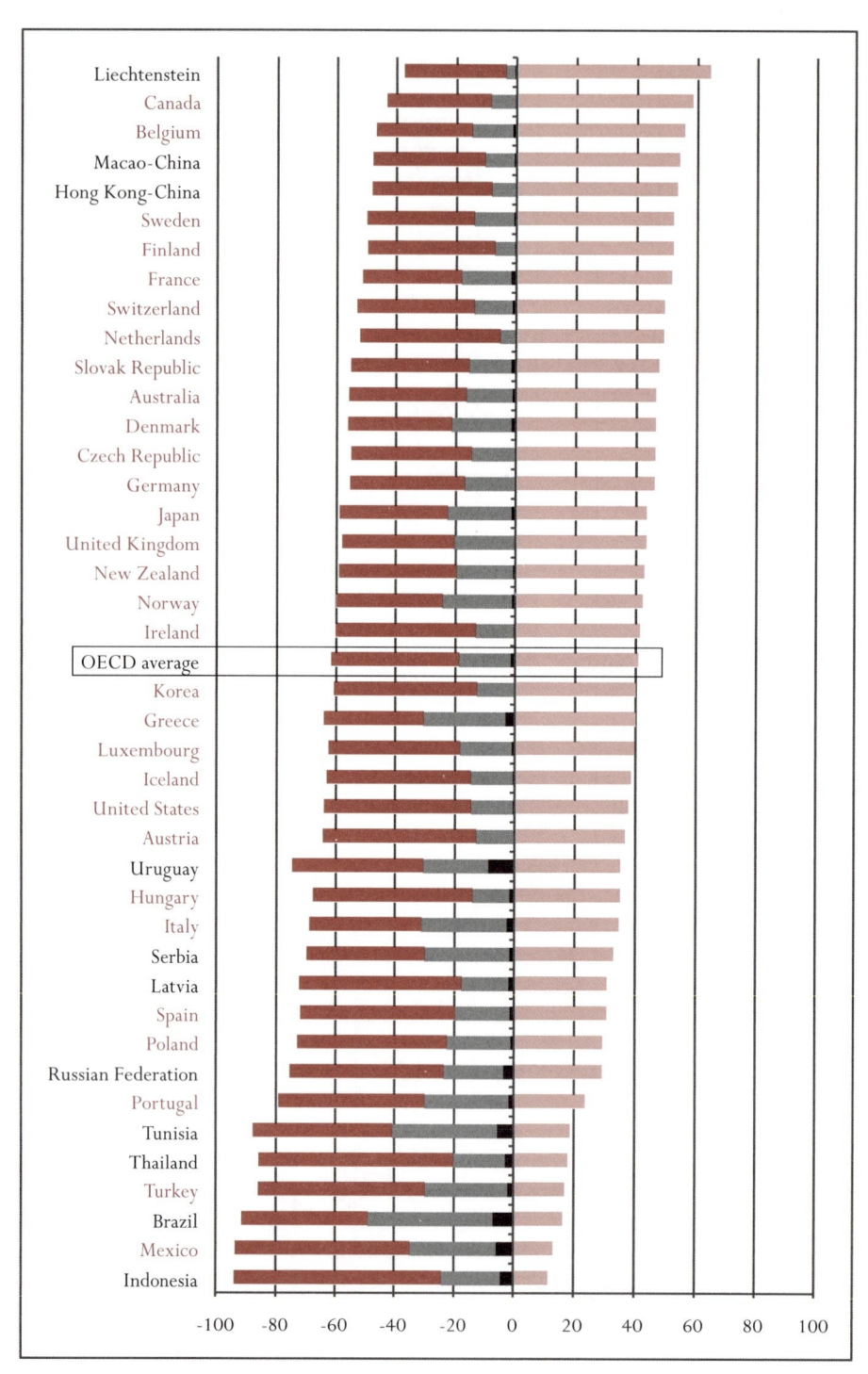

Figure A1.23 ■ **Student performance on Exchange Rate – Question 3**

Percentage of students who:
- Gave an incorrect answer
- Did not reach the question
- Did not answer the question
- Answered correctly – "Yes" with adequate explanation

Liechtenstein
Canada
Belgium
Macao-China
Hong Kong-China
Sweden
Finland
France
Switzerland
Netherlands
Slovak Republic
Australia
Denmark
Czech Republic
Germany
Japan
United Kingdom
New Zealand
Norway
Ireland
OECD average
Korea
Greece
Luxembourg
Iceland
United States
Austria
Uruguay
Hungary
Italy
Serbia
Latvia
Spain
Poland
Russian Federation
Portugal
Tunisia
Thailand
Turkey
Brazil
Mexico
Indonesia

Note: Countries are ranked in descending order of students who gave the correct answer.

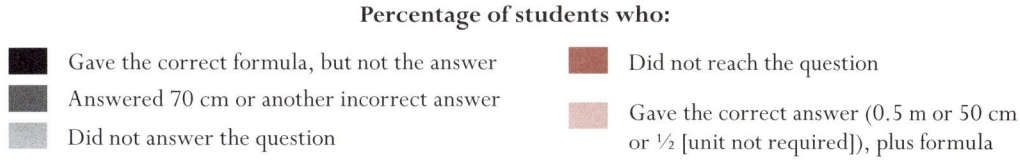

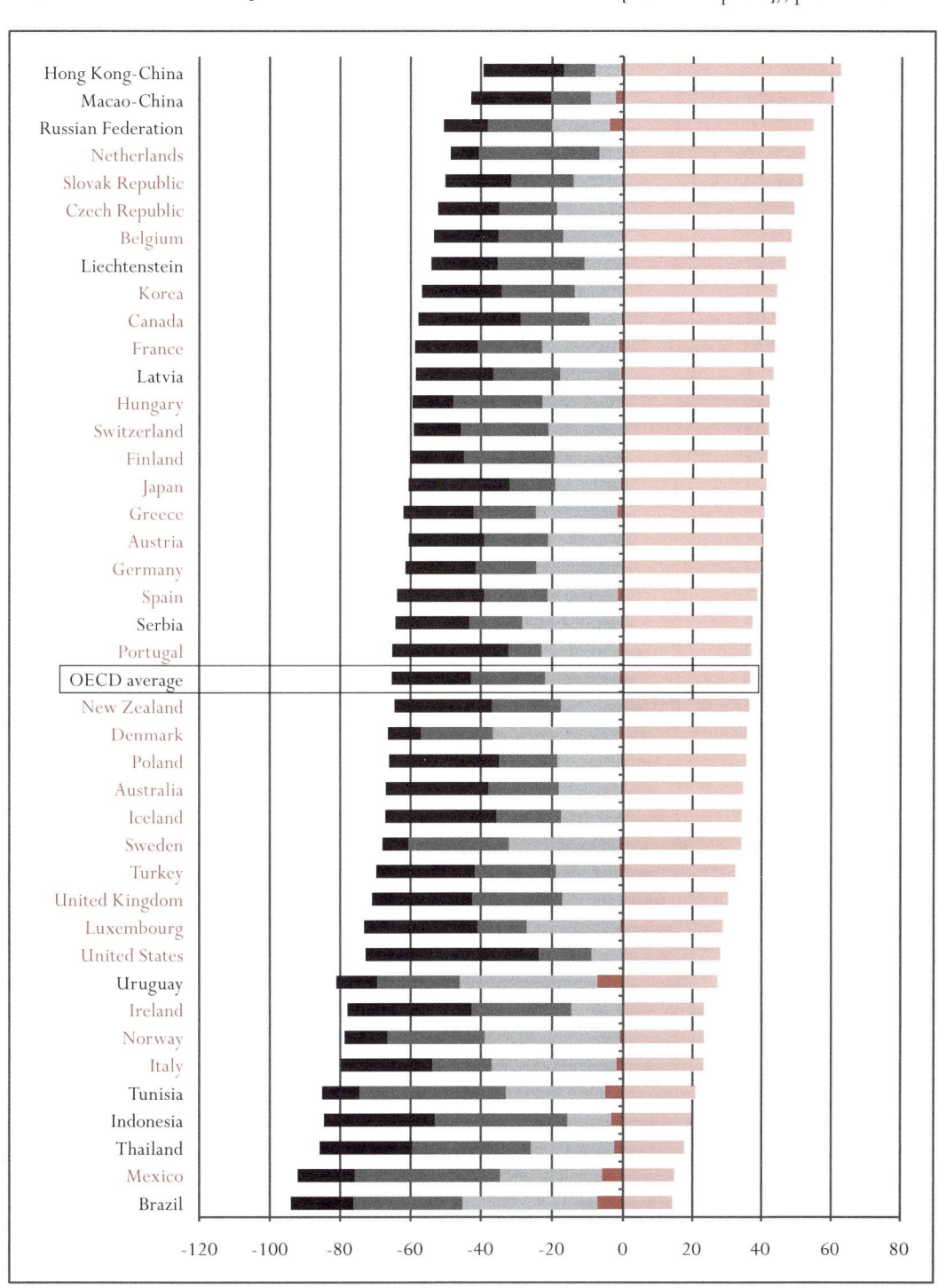

Figure A1.24 ■ **Student performance on P2000 Walking – Question 1**

Percentage of students who:

- Gave the correct formula, but not the answer
- Answered 70 cm or another incorrect answer
- Did not answer the question
- Did not reach the question
- Gave the correct answer (0.5 m or 50 cm or ½ [unit not required]), plus formula

Countries (top to bottom): Hong Kong-China, Macao-China, Russian Federation, Netherlands, Slovak Republic, Czech Republic, Belgium, Liechtenstein, Korea, Canada, France, Latvia, Hungary, Switzerland, Finland, Japan, Greece, Austria, Germany, Spain, Serbia, Portugal, OECD average, New Zealand, Denmark, Poland, Australia, Iceland, Sweden, Turkey, United Kingdom, Luxembourg, United States, Uruguay, Ireland, Norway, Italy, Tunisia, Indonesia, Thailand, Mexico, Brazil

Note: Countries are ranked in descending order of students who gave the correct answer.

Learning Mathematics for Life: A Perspective from PISA – © OECD 2009

Figure A1.25 ■ **Student performance on Support for the President – Question 1**

Percentage of students who:

- Answered "Newspaper 3" but gave no reasons or only one reason
- Answered "Newspaper 4" or other incorrect answer
- Did not answer the question
- Did not reach the question
- Gave the correct answer (Newspaper 3), plus at least two reasons (the poll is more recent, with larger sample size, a random selection of the sample, only voters were asked)

Note: Countries are ranked in descending order of students who gave the correct answer.

Annex A1

Learning Mathematics for Life: A Perspective from PISA – © OECD 2009

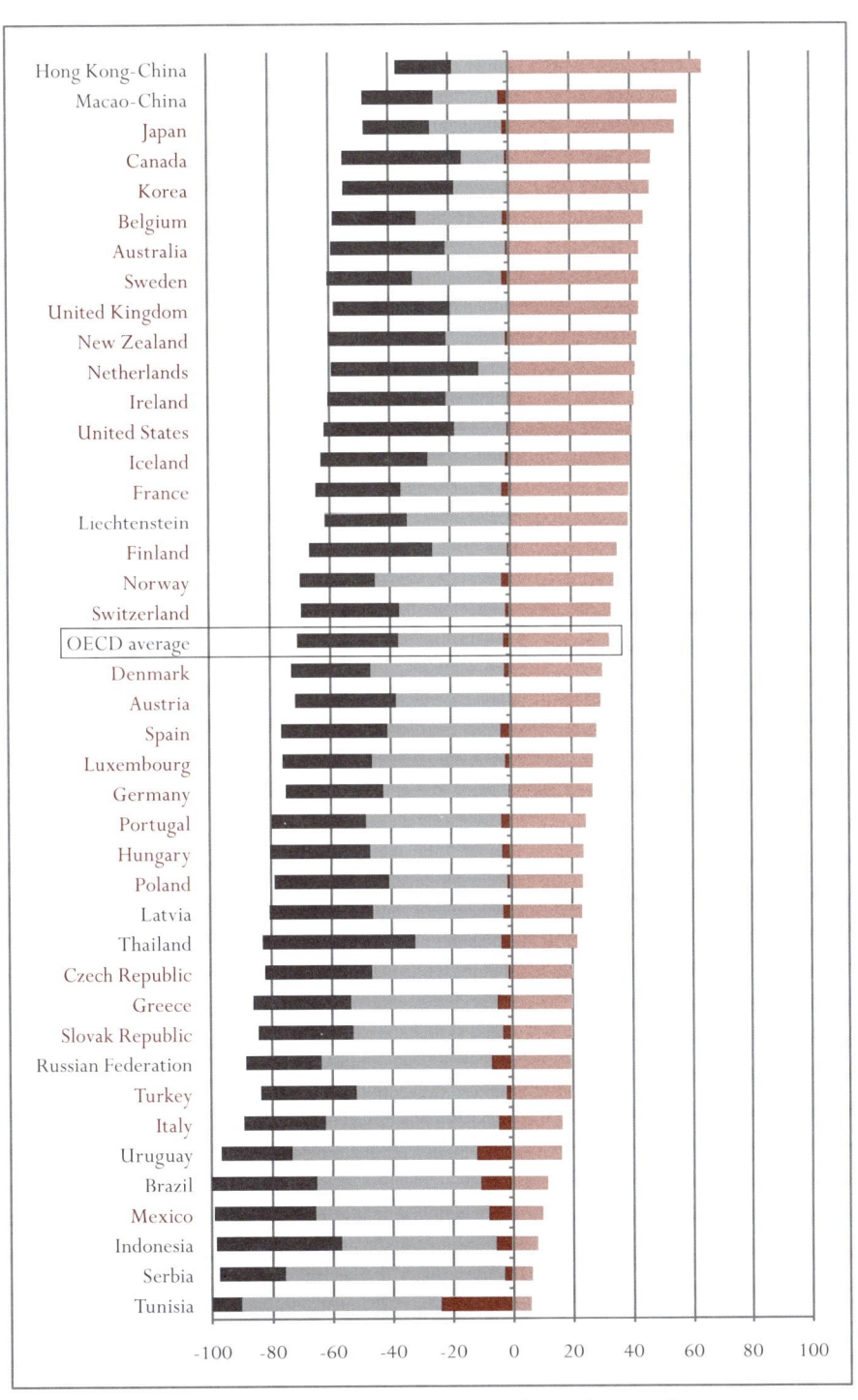

Figure A1.26 ■ **Student performance on Test Scores – Question 1**

Percentage of students who:

- Gave an incorrect answer giving no or wrong mathematical reasons or simply describing differences
- Did not answer the question
- Did not reach the question
- Answered correctly – one valid argument is given (number of students passing, disproportionate influence of outlier, number of students with scores in the highest level)

Note: Countries are ranked in descending order of students who gave the correct answer.

Learning Mathematics for Life: A Perspective from PISA – © OECD 2009

Figure A1.27 ■ **Student performance on Robberies – Question 1**

Percentage of students who:

- Answered "Yes" or other incorrect answer
- Did not answer the question
- Did not reach the question
- Answered correctly – either "No, this focuses on only a small part of the graph" or "Trend data are required"
- Answered "No", but the explanation lacks detail

Note: Countries are ranked in descending order of total percent correct, that is, allowing for both full and partial credits.

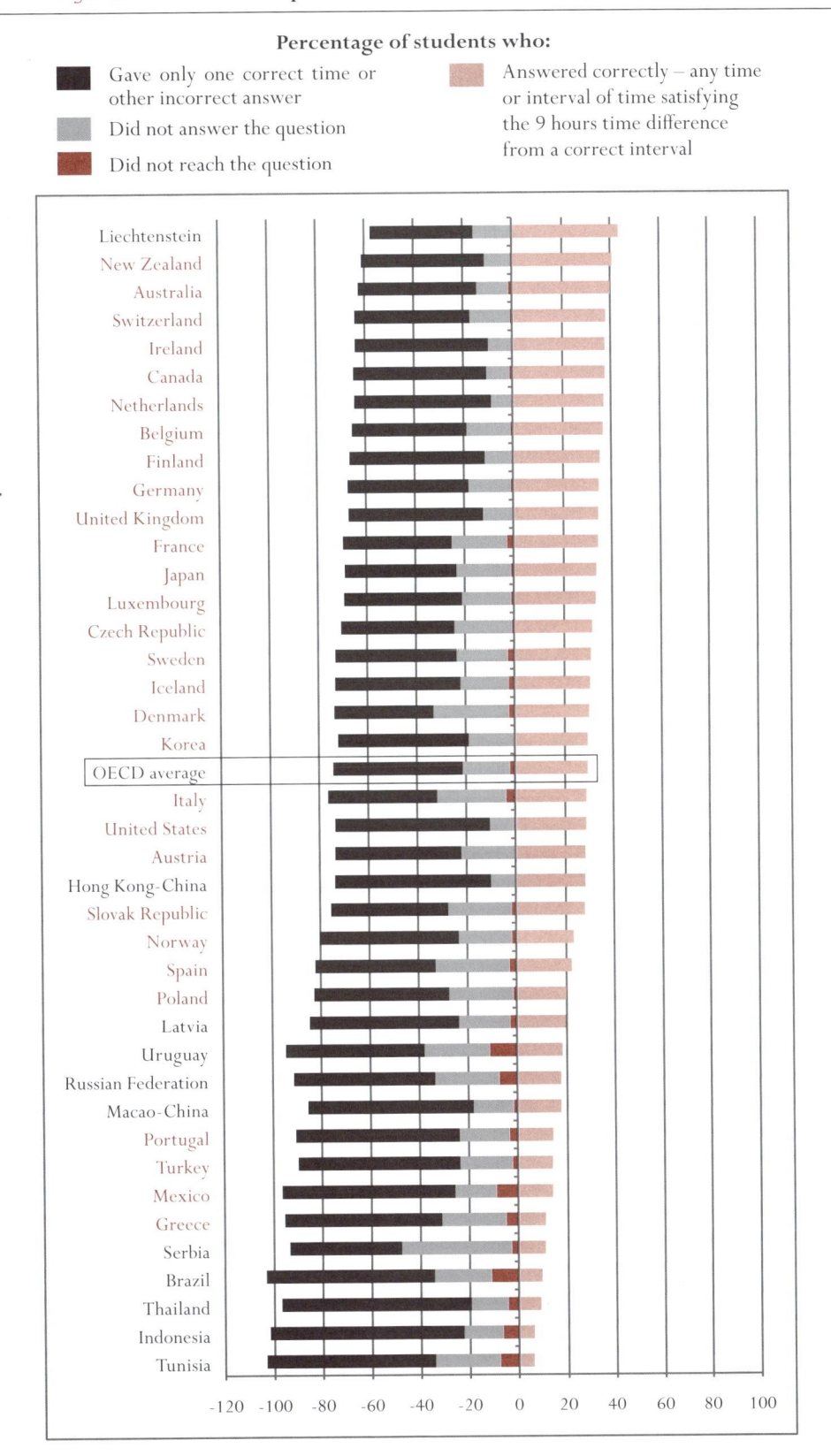

Figure A1.28 ■ **Student performance on Internet Relay Chat – Question 2**

Percentage of students who:
- Gave only one correct time or other incorrect answer
- Did not answer the question
- Did not reach the question
- Answered correctly – any time or interval of time satisfying the 9 hours time difference from a correct interval

Countries (top to bottom): Liechtenstein, New Zealand, Australia, Switzerland, Ireland, Canada, Netherlands, Belgium, Finland, Germany, United Kingdom, France, Japan, Luxembourg, Czech Republic, Sweden, Iceland, Denmark, Korea, OECD average, Italy, United States, Austria, Hong Kong-China, Slovak Republic, Norway, Spain, Poland, Latvia, Uruguay, Russian Federation, Macao-China, Portugal, Turkey, Mexico, Greece, Serbia, Brazil, Thailand, Indonesia, Tunisia.

Note: Countries are ranked in descending order of students who gave the correct answer.

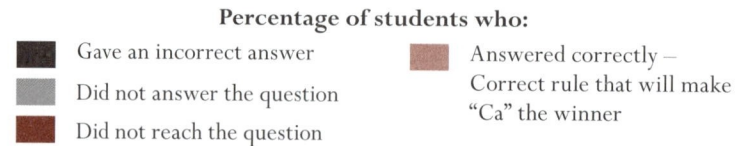

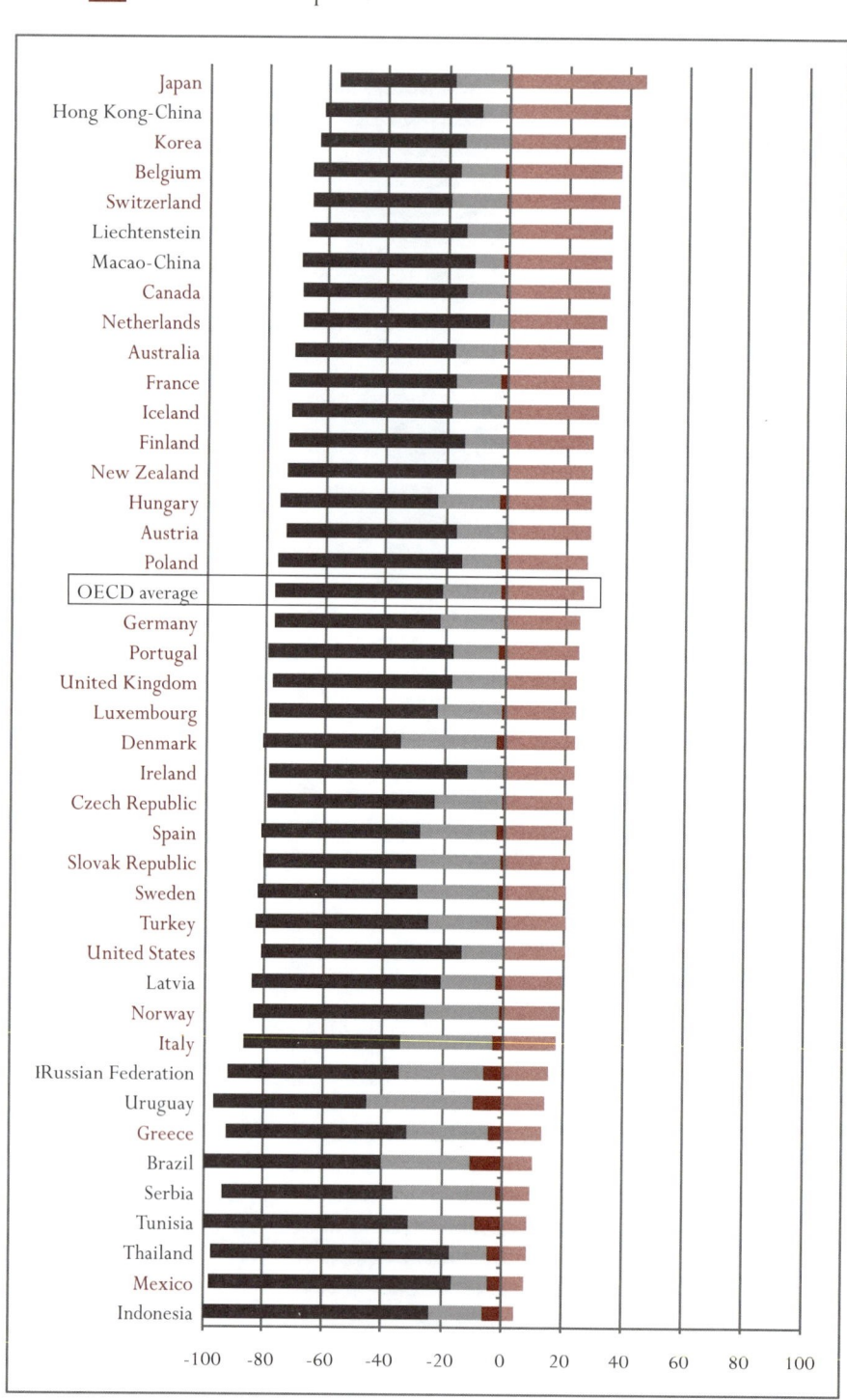

Figure A1.29 ■ **Student performance on The Best Car – Question 2**

Percentage of students who:
- Gave an incorrect answer
- Did not answer the question
- Did not reach the question
- Answered correctly – Correct rule that will make "Ca" the winner

Note: Countries are ranked in descending order of students who gave the correct answer.

Figure A1.30 ■ **Student performance on Walking – Question 3**

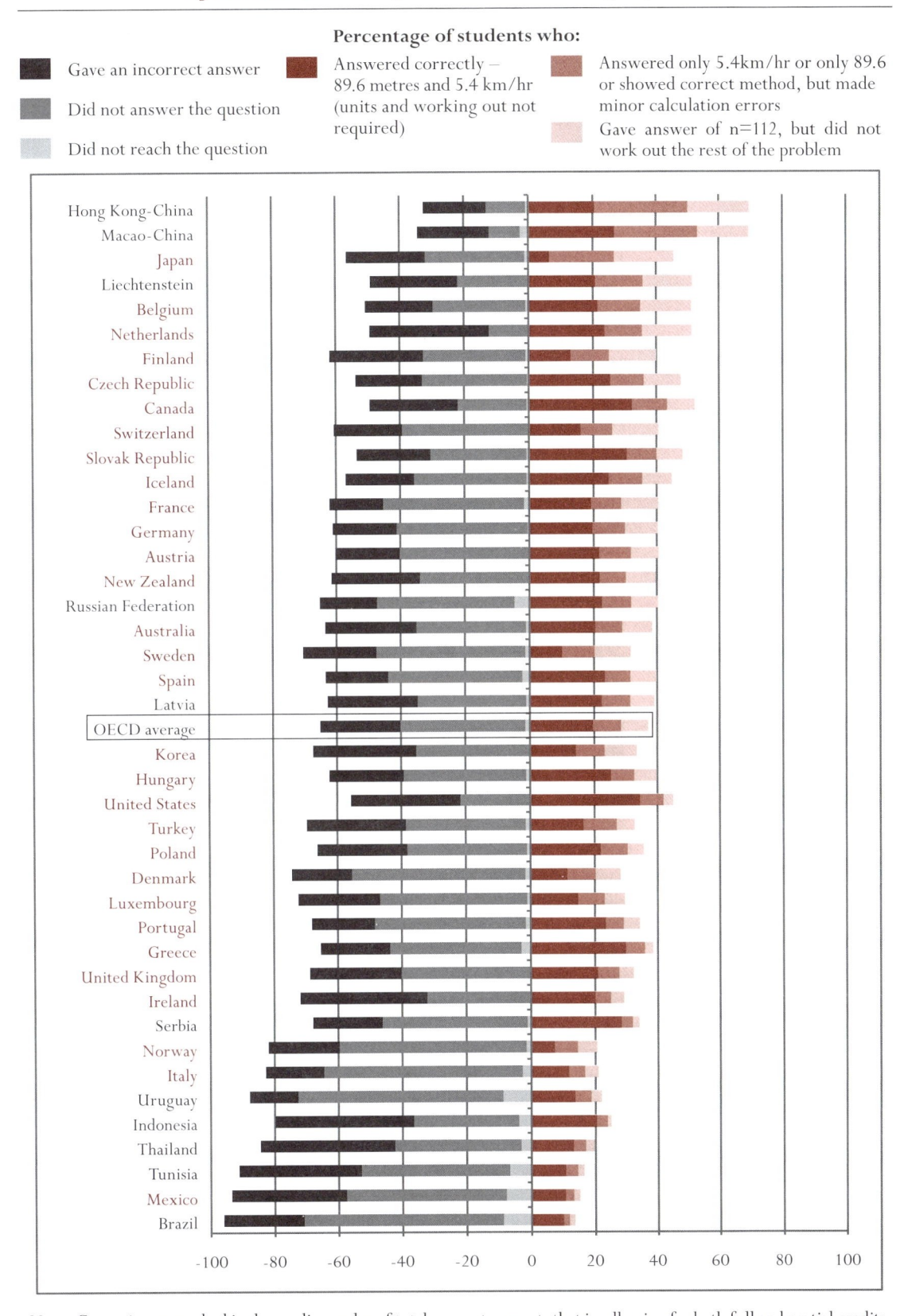

Percentage of students who:

- Gave an incorrect answer
- Did not answer the question
- Did not reach the question
- Answered correctly – 89.6 metres and 5.4 km/hr (units and working out not required)
- Answered only 5.4km/hr or only 89.6 or showed correct method, but made minor calculation errors
- Gave answer of n=112, but did not work out the rest of the problem

Note: Countries are ranked in descending order of total percent correct, that is, allowing for both full and partial credits.

Learning Mathematics for Life: A Perspective from PISA – © OECD 2009

Figure A1.31 ■ **Student performance on Carpenter – Question 1**

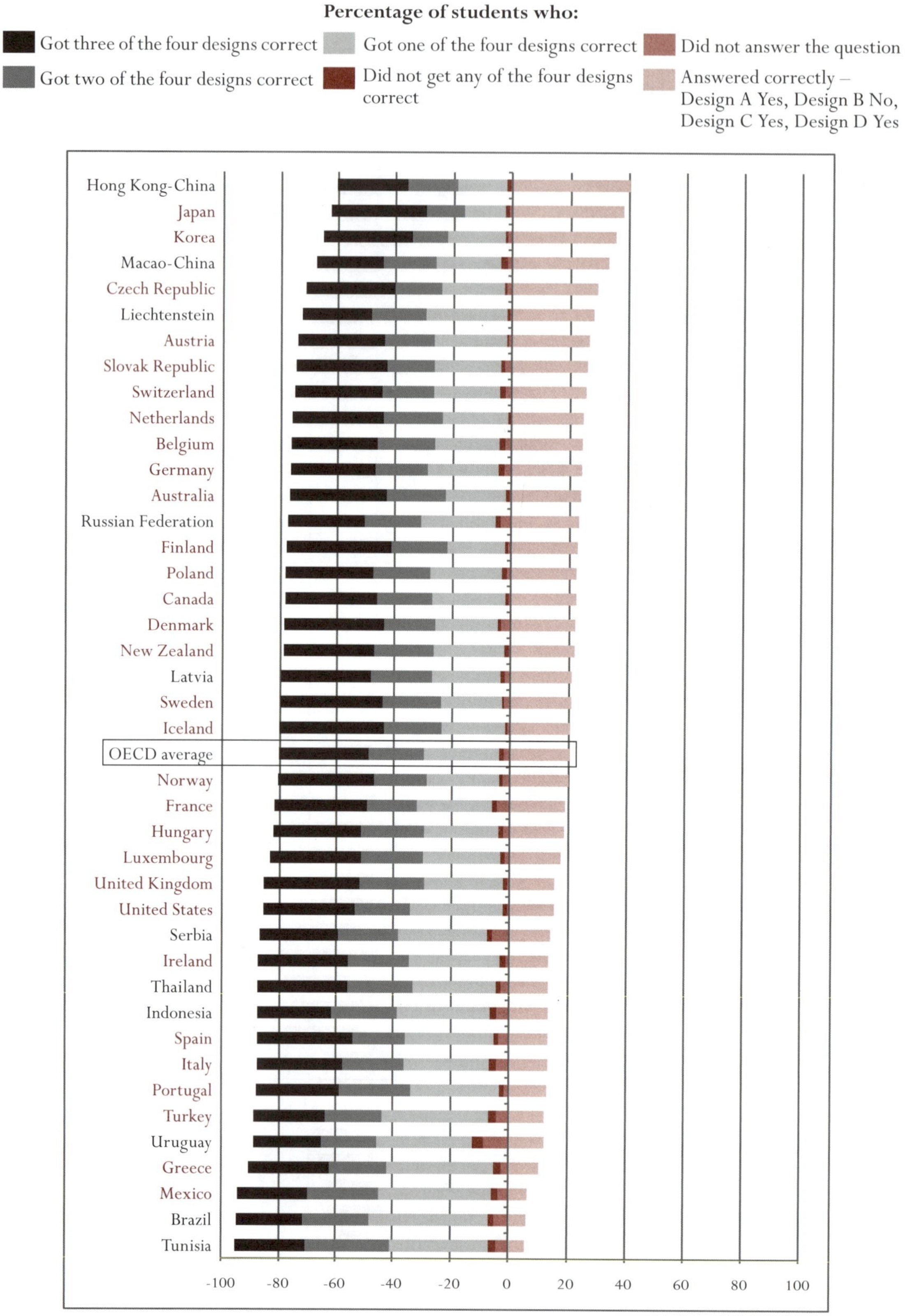
Percentage of students who:
Got three of the four designs correct
Got two of the four designs correct
Got one of the four designs correct
Did not get any of the four designs correct
Did not answer the question
Answered correctly – Design A Yes, Design B No, Design C Yes, Design D Yes
Hong Kong-China
Japan
Korea
Macao-China
Czech Republic
Liechtenstein
Austria
Slovak Republic
Switzerland
Netherlands
Belgium
Germany
Australia
Russian Federation
Finland
Poland
Canada
Denmark
New Zealand
Latvia
Sweden
Iceland
OECD average
Norway
France
Hungary
Luxembourg
United Kingdom
United States
Serbia
Ireland
Thailand
Indonesia
Spain
Italy
Portugal
Turkey
Uruguay
Greece
Mexico
Brazil
Tunisia
-100
-80
-60
-40
-20
0
20
40
60
80
100

Note: Countries are ranked in descending order of students who gave the correct answer.

Annex A2

OTHER EXAMPLES OF PISA MATHEMATICS QUESTIONS THAT WERE NOT USED IN THE PISA 2003 MATHEMATICS ASSESSMENT

This annex includes all PISA mathematics questions that were released before the PISA 2003 main testing was conducted.

Table A2.1
Other examples of released PISA mathematics questions not used in PISA 2003

Item code	Item name	Content area	Competency	Context	Item format
M307Q01	Drug Concentrations: Q1	Change and relationships	Connections	Scientific	Short Answer
M307Q02	Drug Concentrations: Q2	Change and relationships	Connections	Scientific	Multiple Choice
M307Q03	Drug Concentrations: Q3	Change and relationships	Connections	Scientific	Multiple Choice
M309Q01	Building Blocks: Q1	Shape and space	Reproduction	Personal	Short Answer
M309Q02	Building Blocks: Q2	Shape and space	Reproduction	Personal	Short Answer
M309Q03	Building Blocks: Q3	Shape and space	Connections	Personal	Short Answer
M309Q04	Building Blocks: Q4	Shape and space	Reflections	Personal	Short Answer
M432Q01	Reaction Time: Q1	Change and relationships	Reproduction	Scientific	Short Answer
M432Q02	Reaction Time: Q2	Change and relationships	Reflections	Scientific	Extended Response
M465Q01	Water Tank: Q1	Change and relationships	Connections	Scientific	Multiple Choice
M471Q01	Spring Fair: Q1	Uncertainty	Connections	Educational and occupational	Multiple Choice
M472Q01	Swing: Q1	Change and relationships	Connections	Personal	Multiple Choice
M479Q01	Student Heights: Q1	Uncertainty	Reflections	Educational and occupational	Complex Multiple Choice
M480Q01	Payments By Area: Q1	Change and relationships	Connections	Public	Complex Multiple Choice
M480Q02	Payments By Area: Q2	Quantity	Connections	Public	Extended Response
M515Q01	Shoes For Kids: Q1	Change and relationships	Reproduction	Personal	Short Answer
M515Q02	Shoes For Kids: Q2	Change and relationships	Connections	Personal	Multiple Choice
M515Q03	Shoes For Kids: Q3	Change and relationships	Connections	Personal	Multiple Choice
M521Q01	Table Tennis Tournament: Q1	Uncertainty	Reproduction	Personal	Multiple Short Answer
M521Q02	Table Tennis Tournament: Q2	Quantity	Connections	Personal	Multiple Short Answer
M521Q03	Table Tennis Tournament: Q3	Quantity	Connections	Personal	Short Answer
M523Q01	Lighthouse: Q1	Change and relationships	Connections	Public	Multiple Choice
M523Q02	Lighthouse: Q2	Change and relationships	Connections	Public	Multiple Choice
M523Q03	Lighthouse: Q3	Change and relationships	Reflections	Public	Extended Response
M525Q01	Decreasing CO2 Levels: Q1	Quantity	Connections	Scientific	Extended Response
M525Q03	Decreasing CO2 Levels: Q3	Quantity	Reflections	Scientific	Extended Response

Table A2.1
Other examples of released PISA mathematics questions not used in PISA 2003 (continued)

Item code	Item name	Content area	Competency	Context	Item format
M535Q01	Twisted Building: Q1	Shape and space	Connections	Public	Extended Response
M535Q04	Twisted Building: Q4	Shape and space	Connections	Public	Extended Response
M537Q01	Heartbeat: Q1	Change and relationships	Connections	Scientific	Extended Response
M537Q02	Heartbeat: Q2	Change and relationships	Connections	Scientific	Extended Response
M543Q01	Space Flight: Q1	Quantity	Connections	Scientific	Multiple Choice
M543Q02	Space Flight: Q2	Quantity	Connections	Scientific	Multiple Choice
M543Q03	Space Flight: Q3	Quantity	Connections	Scientific	Extended Response
M552Q01	Rock Concert: Q1	Quantity	Connections	Public	Multiple Choice
M703Q01	Moving Walkways: Q1	Change and relationships	Reflections	Scientific	Short Answer
M836Q01	Postal Charges: Q1	Uncertainty	Connections	Public	Multiple Choice
M836Q02	Postal Charges: Q2	Quantity	Connections	Public	Extended Response

Annex A3

TRADITIONAL DOMAINS AND PISA ITEMS

Table A3.1
Traditional domains; average item difficulties (logits) relative to other topics and their standard deviations[1]

Country	Algebra (7 items) Mean (SD)	Data (26 items) Mean (SD)	Geometry (12 items) Mean (SD)	Measurement (8 items) Mean (SD)	Number (32 items) Mean (SD)
Australia	0.94 (1.39)	-0.35 (1.24)	-0.15 (1.18)	0.95 (0.94)	-0.10 (1.06)
Austria	0.72 (1.43)	0.05 (1.33)	-0.20 (1.02)	0.70 (0.95)	-0.29 (1.20)
Belgium	0.75 (1.27)	-0.13 (1.24)	-0.19 (1.15)	0.97 (1.09)	-0.23 (1.01)
Brazil	0.86 (1.82)	-0.17 (1.28)	-0.29 (1.09)	1.50 (1.35)	-0.32 (1.13)
Canada	0.88 (1.43)	-0.37 (1.17)	-0.06 (1.09)	1.16 (0.94)	-0.16 (1.05)
Czech Republic	0.84 (1.36)	0.11 (1.31)	-0.30 (0.97)	0.76 (0.91)	-0.35 (1.21)
Denmark	1.08 (1.45)	-0.25 (1.38)	-0.23 (1.16)	0.92 (1.23)	-0.17 (1.21)
Finland	1.03 (1.31)	-0.28 (1.20)	-0.08 (1.13)	1.04 (0.89)	-0.23 (1.20)
France	0.87 (1.36)	-0.27 (1.28)	-0.10 (1.10)	1.01 (1.16)	-0.19 (1.12)
Germany	0.81 (1.56)	-0.11 (1.27)	-0.16 (1.05)	1.00 (1.15)	-0.28 (1.16)
Greece	0.75 (1.42)	-0.14 (1.11)	-0.29 (1.04)	1.33 (1.06)	-0.28 (1.09)
Hong Kong-China	0.51 (1.28)	-0.11 (0.99)	-0.26 (1.25)	0.82 (0.82)	-0.13 (1.20)
Hungary	0.79 (1.56)	-0.12 (1.27)	-0.26 (1.15)	1.18 (0.82)	-0.27 (1.14)
Iceland	1.13 (1.33)	-0.30 (1.13)	-0.11 (0.95)	1.13 (1.10)	-0.24 (1.08)
Indonesia	0.59 (1.82)	-0.03 (1.15)	-0.31 (1.12)	1.01 (0.99)	-0.24 (1.29)
Ireland	1.17 (1.61)	-0.49 (1.21)	0.13 (1.01)	1.31 (1.20)	-0.23 (1.15)
Italy	1.07 (1.57)	-0.10 (1.27)	-0.12 (1.07)	0.82 (1.05)	-0.31 (1.13)
Japan	0.69 (1.29)	-0.03 (1.24)	-0.40 (1.30)	0.82 (0.84)	-0.19 (1.42)
Korea	0.78 (1.37)	-0.13 (1.18)	-0.30 (1.03)	0.94 (0.85)	-0.18 (1.37)
Latvia	0.84 (1.36)	-0.08 (1.21)	-0.29 (1.03)	0.96 (1.01)	-0.25 (1.16)
Luxembourg	1.03 (1.62)	-0.17 (1.32)	-0.14 (1.04)	1.01 (1.10)	-0.29 (1.07)
Macao-China	0.67 (1.38)	-0.10 (1.14)	-0.18 (1.13)	1.02 (0.84)	-0.25 (1.24)
Mexico	1.08 (1.52)	-0.09 (1.24)	-0.30 (1.13)	1.29 (1.00)	-0.37 (1.11)
Netherlands	0.70 (1.29)	-0.31 (1.23)	-0.07 (1.22)	1.21 (1.01)	-0.18 (1.06)
New Zealand	0.91 (1.45)	-0.35 (1.23)	-0.19 (1.14)	0.96 (1.00)	-0.08 (1.07)
Norway	1.25 (1.28)	-0.36 (1.18)	-0.08 (1.03)	1.21 (1.11)	-0.25 (1.21)
Poland	0.94 (1.53)	-0.18 (1.33)	-0.18 (1.03)	0.97 (0.97)	-0.23 (1.19)
Portugal	0.80 (1.78)	-0.33 (1.44)	-0.10 (1.04)	1.35 (1.08)	-0.20 (1.15)
Russian Fed	0.71 (1.48)	0.14 (1.43)	-0.42 (1.03)	0.98 (1.02)	-0.35 (1.12)
Scotland	0.96 (1.54)	-0.44 (1.26)	0.11 (1.16)	1.08 (0.93)	-0.16 (1.16)
Slovak Republic	0.82 (1.44)	0.19 (1.31)	-0.34 (0.90)	0.84 (0.79)	-0.41 (1.16)
Spain	0.85 (1.47)	-0.20 (1.32)	-0.12 (1.12)	1.30 (1.00)	-0.30 (1.09)
Sweden	1.14 (1.41)	-0.27 (1.22)	-0.09 (1.00)	1.17 (1.21)	-0.29 (1.17)
Switzerland	0.91 (1.37)	-0.04 (1.24)	-0.24 (1.07)	0.73 (0.94)	-0.26 (1.07)
Thailand	1.04 (1.40)	-0.16 (1.26)	-0.36 (1.13)	1.08 (1.22)	-0.23 (1.24)
Tunisia	0.62 (1.59)	0.07 (1.20)	-0.25 (1.03)	0.96 (1.15)	-0.34 (1.16)
Turkey	0.49 (1.35)	-0.25 (1.16)	-0.16 (1.03)	1.34 (1.14)	-0.18 (1.09)
United Kingdom	0.96 (1.46)	-0.50 (1.31)	-0.06 (1.09)	1.20 (1.08)	-0.09 (1.07)
United States	0.89 (1.48)	-0.35 (1.17)	-0.14 (1.08)	1.29 (0.97)	-0.18 (0.98)
Uruguay	0.91 (1.57)	-0.16 (1.23)	-0.11 (1.11)	1.08 (1.06)	-0.29 (0.95)
Serbia	0.91 (1.52)	0.15 (1.27)	-0.15 (0.97)	1.00 (1.06)	-0.51 (1.17)

1. Please note that overall country means are set to 0, therefore, this table does not represent the difficulty of the topic between countries, but only within countries.

Annex A4

WORD-COUNT AND THE 3 Cs – ANALYSIS OF VARIANCE

This annex provides technical details for the analysis of variance performed for the language demand section of Chapter 5.

In all cases a full factorial ANOVA was performed in SPSS using a univariate GLM procedure. The dependent variable in all cases was defined as item difficulty in logits centered at 0 for each country. The two factors in all cases were the same: country and categorized word count (see Chapter 5 for the detailed description of word-count). In the first case no other factors were added (see Table 5.1), in the subsequent cases one of the 3Cs were added (Tables 5.2 through Table 5.5) as a third factor (context, competency and content in the form of traditional topics). The third factor was added to find whether there are interactions between the word-count and features discussed in Chapter 5. Please note that as in Chapter 5, the content is categorized by traditional topics such as Algebra, Data, Geometry, Measurement, and Number.

In addition to the F statistics, that show significance of main effects and interactions, partial $\acute{\eta}^2$ is also provided to assess for the percent of total variance in the dependent variable accounted for by the variance between categories (groups) formed by the independent variable(s).

Table A4.1
Full factorial ANOVA with word-count and country as factors

Source	Type III Sum of Squares	df	Mean Square	F	Sig.	Partial $\acute{\eta}^2$
Word-count	547.805	2	273.902	191.563	0.000	0.105
Country	0.302	39	0.008	0.005	1.000	0.000
Country * Word-count	16.334	78	0.209	0.146	1.000	0.003
Error	4 689.842	3 280	1.430			
Total	5 253.982	3 400				

Table A4.2
Full factorial ANOVA with word-count, country and competencies as factors

Source	Type III Sum of Squares	df	Mean Square	F	Sig.	Partial $\acute{\eta}^2$
Word-count	168.611	2	84.306	80.109	0.000	0.048
Country	0.772	39	0.020	0.019	1.000	0.000
Competency	1 003.119	2	501.560	476.596	0.000	0.230
Competency * Word-count	77.286	4	19.322	18.360	0.000	0.022
Country * Word-count	15.620	78	0.200	0.190	1.000	0.005
Country * Competency	10.138	78	0.130	0.124	1.000	0.003
Error	3 363.403	3 196	1.052			
Total	5 253.982	3 400				

Table A4.3
Full factorial ANOVA with word-count, country and content as factors

Source	Type III Sum of Squares	df	Mean Square	F	Sig.	Partial $\acute{\eta}^2$
Word-count	111.663	2	55.832	49.443	0.000	0.031
Country	5.689	39	0.146	0.129	1.000	0.002
Content	650.356	4	162.589	143.983	0.000	0.156
Content * Word-count	353.445	7	50.492	44.714	0.000	0.091
Country * Word-count	13.851	78	0.178	0.157	1.000	0.004
Content * Country	60.126	156	0.385	0.341	1.000	0.017
Error	3 515.263	3 113	1.129			
Total	5 253.982	3 400				

Table A4.4
Full factorial ANOVA with word-count, country and context as factors

Source	Type III Sum of Squares	df	Mean Square	F	Sig.	Partial η^2
Word-count	569.862	2	284.931	208.467	0.000	0.117
Country	0.471	39	0.012	0.009	1.000	0.000
Context	167.604	3	55.868	40.875	0.000	0.037
Context * Word-count	183.411	6	30.568	22.365	0.000	0.041
Country * Word-count	12.781	78	0.164	0.120	1.000	0.003
Context * Country	20.949	117	0.179	0.131	1.000	0.005
Error	4 310.861	3 154	1.367			
Total	5 253.982	3 400				

Table A4.5
Post hoc comparisons for word-count mean difficulties using Bonferroni adjustment

Word Count (I)	Word Count (J)	Mean Difference (I-J)	Std. Error	Sig.	99% Confidence Interval Lower Bound	99% Confidence Interval Upper Bound
Short	Medium	0.00828	0.052187	1.000	-0.14502	0.16157
	Long	-0.84144	0.054174	0.000	-1.00057	-0.68231
Medium	Short	-0.00828	0.052187	1.000	-0.16157	0.14502
	Long	-0.84972	0.047475	0.000	-0.98917	-0.71026
Long	Short	0.84144	0.054174	0.000	0.68231	1.00057
	Medium	0.84972	0.047475	0.000	0.71026	0.98917

Annex A5

ANALYSIS OF VARIANCE RELATED TO ITEM FORMAT

This annex provides technical details for the analysis of variance performed for the item format section of Chapter 5.

In all cases a Full Factorial ANOVA was performed in SPSS using a univariate GLM procedure. The dependent variable in all cases was defined as item difficulty in logits centered at 0 for each country. The two factors in all cases were the same: country and item-format. In the first case no other factors were added (see Table A5.1), in the subsequent cases one of the following factors were added (Table A5.2-Competency, Table A5.3-Context, and Table A5.4-Word-Count). Table A5.5 presents the distribution of items by item format and traditional topic groupings. Please note that interactions between topics and item format were not considered due to the very small number of items in each cell (see Table A5.5). Table A5.6 presents *post hoc* comparisons for item format mean difficulties using the Bonferroni adjustment for multiple comparisons.

In addition to the F statistics, that shows significance of main effects and interactions, partial $\acute{\eta}^2$ is also provided to assess for the percent of total variance in the dependent variable accounted for by the variance between categories (groups) formed by the independent variable(s).

Table A5.1
Full factorial ANOVA with item-format and country as factors

Source	Type III Sum of Squares	df	Mean Square	F	Sig.	Partial $\dot{\eta}^2$
Format	626.798	4	156.700	109.065	0.000	0.120
Country	1.605	39	0.041	0.029	1.000	0.000
Country * Format	29.583	156	0.190	0.132	1.000	0.006
Error	4 597.600	3 200	1.437			
Total	5 253.982	3 400				

Table A5.2
Full factorial ANOVA with item-format, country and competencies as factors

Source	Type III Sum of Squares	df	Mean Square	F	Sig.	Partial $\dot{\eta}^2$
Format	271.631	4	67.908	65.563	0.000	0.078
Country	1.551	39	0.040	0.038	1.000	0.000
Competency	770.719	2	385.360	372.053	0.000	0.193
Format * Competency	113.499	7	16.214	15.654	0.000	0.034
Country * Format	28.953	156	0.186	0.179	1.000	0.009
Country * Competency	10.222	78	0.131	0.127	1.000	0.003
Error	3 224.334	3 113	1.036			
Total	5 253.982	3 400				

Table A5.3
Full factorial ANOVA with item-format, country and context as factors

Source	Type III Sum of Squares	df	Mean Square	F	Sig.	Partial $\dot{\eta}^2$
Format	441.548	4	110.387	86.526	0.000	0.101
Country	1.957	39	0.050	0.039	1.000	0.000
Context	91.389	3	30.463	23.878	0.000	0.023
Context * Format	559.524	11	50.866	39.871	0.000	0.125
Country* Format	30.789	156	0.197	0.155	1.000	0.008
Context * Country	25.709	117	0.220	0.172	1.000	0.007
Error	3 915.339	3 069	1.276			
Total	5 253.982	3 400				

Table A5.4
Full factorial ANOVA with item-format, country and word-count as factors

Source	Type III Sum of Squares	df	Mean Square	F	Sig.	Partial η^2
Format	466.839	4	116.710	88.561	0.000	0.102
Country	1.898	39	0.049	0.037	1.000	0.000
Word-count	256.207	2	128.104	97.207	0.000	0.059
Word-count * Format	96.723	6	16.120	12.232	0.000	0.023
Country * Format	28.778	156	0.184	0.140	1.000	0.007
Word-count * Country	15.529	78	0.199	0.151	1.000	0.004
Error	4 103.759	3 114	1.318			
Total	5 253.982	3 400				

Table A5.5
Item distribution across item-format categories and traditional topics

Topic	Complex Multiple Choice	Multiple Choice	Multiple Short Answer	Open Constructed Response	Short Answer	Total
Algebra	1	1	1	2	2	7
Data	3	5	1	9	8	26
Geometry	3	3	0	0	6	12
Measurement	1	1	0	1	5	8
Number	3	8	3	2	16	32
Total	11	18	5	14	37	85

Table A5.6
Post hoc comparisons for item format mean difficulties using Bonferroni adjustment

Item Format (I)	Item Format (J)	Mean Difference (I-J)	Std. Error	Sig.	99% Confidence Interval Lower Bound	99% Confidence Interval Upper Bound
Complex Multiple Choice	Multiple Choice	0.91573	0.072532	0.000	0.67684	1.15461
	Multiple Short Answer	0.36958	0.102221	0.003	0.03291	0.70625
	Open Constructed Response	-0.36749	0.076361	0.000	-0.61899	-0.11599
	Short Answer	0.55356	0.065085	0.000	0.33920	0.76793
Multiple Choice	Complex Multiple Choice	-0.91573	0.072532	0.000	-1.15461	-0.67684
	Multiple Short Answer	-0.54614	0.095808	0.000	-0.86169	-0.23059
	Open Constructed Response	-1.28322	0.067536	0.000	-1.50565	-1.06078
	Short Answer	-0.36216	0.054463	0.000	-0.54154	-0.18278
Multiple Short Answer	Complex Multiple Choice	-0.36958	0.102221	0.003	-0.70625	-0.03291
	Multiple Choice	0.54614	0.095808	0.000	0.23059	0.86169
	Open Constructed Response	-0.73707	0.098739	0.000	-1.06228	-0.41187
	Short Answer	0.18398	0.090302	0.417	-0.11344	0.48140
Open Constructed Response	Complex Multiple Choice	0.36749	0.076361	0.000	0.11599	0.61899
	Multiple Choice	1.28322	0.067536	0.000	1.06078	1.50565
	Multiple Short Answer	0.73707	0.098739	0.000	0.41187	1.06228
	SA Short Answer	0.92105	0.059468	0.000	0.72519	1.11692
Short Answer	Complex Multiple Choice	-0.55356	0.065085	0.000	-0.76793	-0.33920
	Multiple Choice	0.36216	0.054463	0.000	0.18278	0.54154
	Multiple Short Answer	-0.18398	0.090302	0.417	-0.48140	0.11344
	Open Constructed Response	-0.92105	0.059468	0.000	-1.11692	-0.72519

Annex A6

Mathematics Expert Group

Jan de Lange, Chair
Utrecht University
The Netherlands

Zbigniew Marciniak
Warsaw University
Poland

Werner Blum, Chair
University of Kassel
Germany

Mogens Niss
Roskilde University
Denmark

Vladimir Burjan
National Institute for Education
Slovak Republic

Kyung-Mee Park
Hongik University
Korea

Sean Close
St Patrick's College
Ireland

Luis Rico
University of Granada
Spain

John Dossey
Consultant
United States of America

Yoshinori Shimizu
Tokyo Gakugei University
Japan

Mary Lindquist
Columbus State University
United States of America

OECD PUBLISHING, 2, rue André-Pascal, 75775 PARIS CEDEX 16
PRINTED IN FRANCE
(98 2009 11 1 P) ISBN 978-92-64-07499-6 – No. 57051 2010